D0126200

Assessing
Higher Order Thinking
in Mathematics

Edited by

Gerald Kulm

Director, Science, Mathematics, and Technology Programs
AAAS Directorate for Education and Human Resources

AMERICAN ASSOCIATION FOR THE ADVANCEMENT OF SCIENCE

1333 H Street, NW, Washington, DC 20005

Library of Congress Cataloging-in-Publication Data

Kulm, Gerald
    Assessing higher order thinking in mathematics / edited by Gerald Kulm
        p.   cm.
    ISBN 0-87168-356-3
    1. Mathematics — Study and teaching — United States.   2. Mathematical ability — Testing.
I. Kulm, Gerald

QA13.A877   1990
510′.71′073 — dc20   90-572
        CIP

Second printing 1991

Publication No. 89-27S

© 1990 by the American Association for the Advancement of Science
1333 H Street, NW, Washington, DC 20005

Acknowledgments

I would like to thank the members of the AAAS staff who contributed to making this book possible. Kathryn Wolff provided support for the proposal and continued to help in many ways in its production. Sue O'Connell was proficient, patient, and creative in editing and preparing the manuscript for production. Susan Cherry and Elisabeth Carroll prepared figures and did final production work; Joan Levy designed the cover. Finally, I am especially grateful to Danita Wilkinson who tracked manuscripts, did all of the computer entry work, and helped produce the index. While I am responsible for the final content, the book is a reality because of these dedicated people.

Gerald Kulm

Cover art modified from Mark Wilson, "Investigating Structured Problem-Solving Items," this volume.

The views presented in this publication are those of the authors and do not necessarily represent those of the Board of Directors or the membership of the American Association for the Advancement of Science.

To JOAN

Contents

Assessing Higher Order Mathematical Thinking: What We Need to Know and Be Able to Do

GERALD KULM

Most mathematics teachers are able to construct a test for a specific class or course in order to assess students' attainment of a set of objectives, perhaps including objectives that could be identified as higher order thinking. Furthermore, many teachers would probably use, more or less informally, a framework such as cognitive level by mathematical content to establish the breadth and depth of items included in the test. That is, the teacher would use his or her own knowledge about the abilities of the students in the class to construct or choose a set of items that include a few easy ones, a majority that are moderately difficult, and a few challenging ones. The teacher would also make sure that all of the mathematical content that is deemed important is covered by the test items.

Unless our notions about assessment are likely to change significantly, this type of a procedure will continue to be used by teachers and test developers. More importantly, this approach to testing is likely to remain as the primary criterion that mathematics teachers, school administrators, and the general public use to define what mathematics is and how well students have learned. In order to make changes in school mathematics, we must be able to specify the parameters that determine how much and what part of a subject has been learned by whom. One approach is to modify the traditional cognitive-level-by-content

framework so that it reflects new conceptions of mathematics thinking and learning. Another approach—embraced by those who believe that a two-dimensional framework cannot possibly reflect the way mathematics is learned—is to start anew, building upon more recent psychological models of learning and more modern ways of looking at the content of mathematics.

A second fundamental issue for assessment concerns the question: What is mathematics? In assessing higher order thinking, we must be able to identify the parameters that characterize mathematical thinking processes. In the case of applications and other problem-based contexts for doing mathematics, we need to be able to identify essential mathematical content that is embedded and to have some idea about how the context and content interact with performance.

Mathematics has often been chosen as an example subject area for experiments by educational psychologists and test developers, partly because it appears to be so readily amenable to breaking into nice, simple, linear pieces. It might be argued that the behavioral objectives craze and the focus on learning hierarchies were largely responsible for widening the gap between school mathematics and "real" mathematics. Anything that could not be stated and measured behaviorally gradually disappeared from the curriculum. Is real mathematics compatible with the kind of assessment currently required by our society?

An issue left over from previous mathematics reforms is the distinction between pure and applied mathematics. One unintended legacy of the modern math movement of the 1960s is the notion that one should first learn the underlying mathematics, then apply the principles. Even the problem-solving emphasis of the 1980s has not changed textbooks much in this respect. In most cases, computational skills are presented, practiced, then applied in story problems at the end of the chapter.

Ideally, now that applications of mathematics are again acceptable as part of the curriculum, the boundary between mathematics and other fields will begin to blur. But distinctions remain. One view, as defined by documents such as the *Curriculum and Evaluation Standards for School Mathematics* (NCTM, 1989), is that mathematics is the central focus, characterized by its power in modeling and communicating the concepts and relationships in the sciences and technology. Another perspective, offered by reports such as *Science for All Americans* (AAAS, 1989), is that mathematics is a tool of science and technology, essential for carrying out and describing the processes and outcomes of theoretical and empirical investigations. Obviously, both of these views, as well as the positions between them, are important. But it is less clear how these perspectives of mathematics can be balanced to produce a more effective approach to school mathematics curriculum and instruction. Assessment will play a major role as this issue evolves. If the mathematics education community retreats once again to easily measured areas of mathematical performance, the richness of applications and pure mathematical thought—which has essentially disappeared from school mathematics over the past 25 years—will not be recovered.

A final aspect of assessment involves determining the age or mathematical background of the students for whom an item or test is appropriate. With some notable exceptions (e.g., pretests and aptitude tests), we generally wait until students are expected to have had the opportunity to learn most of the material before testing. Another interesting exception to this practice is some of the recent research on mathematical problem solving. In order to make sure that a problem is really a problem—that is, the student has no algorithm or standard method readily available—many researchers have used mathematical content that is a year or two ahead of the students' mathematical experience. A typical approach is to use problems easily solved algebraically for students who have not yet studied algebra. This research has shown that the issue of who has learned what mathematics at a particular age is a complex matter. Students who are given a chance to be inventive, working individually or in groups with access to aids and without time constraints, are likely to do surprising things. Given encouragement, or simply a nonjudgmental environment, they often persist and eventually solve problems that they would not encounter in their mathematics classrooms. Furthermore, students in this environment are very likely to use strategies unlike those usually taught in mathematics classes (Kulm, 1982). On the other hand, when placed in problem-solving situations, students with fairly strong mathematics training may perform below expectations. When they are removed from comfortable settings in which a test can be expected to include well-studied approaches and algorithms, even good students stumble, resorting to guesswork and following blind alleys (Schoenfeld, 1985). Clear notions of students' preparation and abilities—as well as how these interact with the test setting—are essential for developing valid higher order thinking assessment approaches.

These fundamental notions about assessment will now be discussed in more detail. The discussion is structured around some specific statements about the things we need to know and be able to do. The statements are not given in any particular order or level of priority.

We need to be able to construct frameworks, descriptions, and specifications for mathematical content, processes, skills, abilities, and performances. In many large assessment efforts such as the National Assessment of Educational Progress (NAEP), a panel of experts attempts to construct a usable framework on which to build a set of items. Recently, there has been movement away from traditional process-by-content frameworks, which are often inadequate for the task of constructing items that address problem solving and other higher order thinking. In a NAEP project that developed and pilot tested prototype items to assess science and mathematics thinking (Blumberg et al., 1985), the advisory group did eventually propose and describe a new framework. However, it became clear during the test development phase that the framework was not very helpful as a guide for constructing items. In a later phase, when the group attempted to interpret the results of student responses to the items, the value of such a framework once again became clear. Without a good framework, there was little guidance for deciding whether items were good or bad, whether student responses to the items were adequate or not, or indeed which problem-solving abilities the students did or did not possess (Kulm, 1986).

We need to be able to define and categorize fundamental higher order thinking processes in mathematics, as well as other important processes that have specific applications in mathematical contexts. In the NAEP higher order assessment project, the advisory panel worried about including mathematical content that went beyond basic arithmetical operations on whole numbers, even for eleventh graders. If students could not do an item, how could one know whether it was because the mathematics was unfamiliar or because a particular higher order thinking process was not present? Furthermore, it was not entirely clear beyond computational procedures what constituted a mathematical process. Is a process mathematical because it is applied in a mathematical context? What are the mathematical processes other than the computational ones? Without clear answers to these questions, the resulting test items were sometimes difficult to classify as mathematical. In interpreting the responses by students, it was sometimes unclear whether the solutions reflected knowledge or processes that might have been learned about mathematics or whether the solutions were developed through the student's general reasoning skills. This kind of issue has some fundamental implications for deciding what should go on in a mathematics class.

We need to be able to specify the mathematical thinking processes and abilities expected of students at key points in their mathematical development and education. In the NAEP project, the advisory panel was faced with constructing problem items for third, eighth, and eleventh graders. Its members chose to minimize prerequisite mathematical knowledge in order to focus on thinking processes. But how much do we know about the higher order thinking processes that are employed at different ages or developmental levels? Much of the research on mathematical problem solving has described common processes (e.g., counting, guessing, drawing pictures) that individual students use on many different types of problems. Often students are more likely to use these informal and intuitive strategies than the more formal approaches learned in mathematics classes. Furthermore, these informal strategies are used by students of widely varying ages and mathematical experience, making it difficult to distinguish, except in very broad categories (e.g., expert versus novice), among those with different mathematical backgrounds. This is particularly true on nonroutine or applied problems (Kulm, 1982). Even in the case of simple addition and subtraction, young children who see the items as problems rather than as exercises use invented counting strategies and other intuitive approaches (Carpenter, 1985). Scope and sequence charts, course outlines, or instructional units that include problem-solving objectives are usually based mainly on intuition, experience, and guesswork rather than on research about children's capabilities at particular ages or mathematical backgrounds. The result is that assessments of student learning do not make it possible to gauge the growth of problem-solving abilities.

As noted by Glaser (1986) at a conference on the redesign of testing, the activities of instruction and testing are likely to become even more closely intertwined in the future. Testing will begin to have a different and more direct relation to instruction. Up until now, major efforts and advances have been made in aptitude tests to predict possible success and standardized tests to denote standards and assess competence upon the completion of instruction. In the future, tests and other assessment tools will be used more often to estimate development toward competence and to guide the instructional process. This change will place a greater reliance upon our knowledge about the learning process in the construction of tests.

In the case of higher order mathematical thinking, early efforts to assess children's performance in clinical settings have been very much in line with the notion of measuring progress and development. Some attempts have been made to use more standardized approaches to problem-solving assessment that are in line with the traditional purposes of testing (Charles et al., 1984; Schoen & Oemke, 1979). However, efforts such as these appear to have had little direct impact on instruction. We need to do more to help teachers use the methods developed through clinical studies. In England, the Assessment Performance Unit (APU) has consciously used its testing program as a tool to develop teachers' skills in assessing higher order thinking (Murphy & Schofield, 1984). There is hope and some anecdotal evidence from APU that involving teachers in the development and administration of higher order thinking assessment programs encourages them to use new instructional strategies. We have already begun to see that little progress can be made in teaching problem solving and other types of higher order mathematical thinking without a concomitant effort to align curriculum with instructional improvement efforts. That alignment is unlikely to be possible without significant advances in the conception and design of approaches to assessment.

References

American Association for the Advancement of Science (1989). *Science for all Americans.* Washington, DC: AAAS.

Blumberg, F., Epstein, M., & MacDonald, W. (1985). *National Assessment of Educational Progress higher order skills planning conference.* Princeton, NJ: National Assessment of Educational Progress.

Carpenter, T. P. (1985). Learning to add and subtract: An exercise in problem solving. In E. A. Silver (Ed.) *Teaching and learning mathematical problem solving: Multiple research perspectives.* Hillsdale, NJ: Lawrence Erlbaum.

Charles, R., Lester, F., & O'Daffer, P. (1984). *An assessment model for problem solving.* Springfield, IL: Illinois State Board of Education.

Glaser, R. (1986). The integration of instruction and testing. In *The redesign of testing for the 21st century.* Princeton, NJ: Educational Testing Service.

Kulm, G. (1982). *Analysis and synthesis of mathematical problem solving processes.* Final Report, Grant #SED-79-20596, Purdue University, West Lafayette, IN.

Kulm, G. (1986). Assessing higher order thinking in mathematics. Paper presented at the NCTM RAC/SIG Research Presession, Washington, DC.

Murphy, P. & Schofield, B. (1984). *Assessment of Performance Unit science report for teachers: Science at age 13.* Department of Education and Science, Welsh Office; Department of Education for Northern Ireland.

National Assessment of Educational Progress. (1983). *The third national mathematics assessment: Results, trends, and issues.* Denver: National Assessment of Educational Progress.

National Council of Teachers of Mathematics. (1989). *Curriculum and evaluation standards for school mathematics.* Reston, VA: National Council of Teachers of Mathematics.

Schoen, H. L. & Oehmke, T. M. (1979). *The IPSP problem-solving test.* Cedar Falls, IA: University of Northern Iowa.

Schoenfeld, A. H. (1985). *Mathematical problem solving.* New York: Academic Press.

I

Current Perspectives on Mathematics Assessment

1

Developing Comprehensive Assessments of Higher Order Thinking

EVA L. BAKER

Testing and related achievement assessment methods undeniably represent one of the most widespread and powerful approaches that are being used to control the quality of schooling. It is also clear that the impact of tests in the service of accountability is not unbridled good. Critics contend that achievement tests may be shallow, corruptible, or incorrect (Shepard, 1989; Linn, Graue, & Sanders, 1989; Burstein, 1989; Baker, 1989; Koretz, 1989). This chapter will attempt to place in context the renewed attention to assessments that attempt to capture more complex aspects of educational attainments of students.

How Is Higher Order Thinking Conceived and Measured?

Higher order thinking has been conceived both in terms of analyses of intellectual processes and of task characteristics.

Intellectual attributes of higher order thinking

In simplest terms, higher order thinking measures include all intellectual tasks that call for more than information retrieval. Any transformation of information is definitionally "higher order" thinking. Early writers who took the approach of detailing intellectual processes and illustrative tasks included

Bloom (1956) and Gagne (1985). Operating from an assessment perspective, Bloom and his colleagues formulated an analysis that popularized the term "higher order," since the upper levels of their venerable *Taxonomy of Educational Objectives* (comprehension, application, analysis, synthesis, and evaluation) provided an operational description that influenced test developers and curriculum designers for years. Gagne's analysis of intellectual processes required by tasks, developed from a learning and training perspective, was of similar influence and was also frequently construed to have a hierarchical character. Both analyses rest upon the inference of transformation and construction processes from tasks, and have served as important precursors to many current cognitive analyses.

Other formulations of higher order thinking derive from the general problem-solving literature and emphasize such task components as problem identification and solution testing. These may treat problem solving either as a subject-matter domain independent task, akin to general critical thinking (Ennis, 1987), or as dependent upon particular content domains. Higher order thinking can also take the form of metacognitive skills, such as planning and self-checking. These skills may be either independent of or embedded in the subject matter task to which they are applied. Other formulations focus on intellectual processes as they are unique to particular subject matter domains (e.g., the use of appropriate rhetorical structure in written composition).

Higher order assessment tasks

On the assessment plane, it is common to ascribe higher order thinking to relatively global and surface features of the tasks themselves. Certain task attributes are thought to require higher order processing. For example, "open-ended" questions describe tasks for which many answers are appropriate and imply that higher order processes are necessary to respond (California State Department of Education, 1989). Obviously, open-ended questions can also solicit a range of information retrieval (e.g., tell all the facts you know about dinosaurs). In the same way, almost all student constructions, such as those included in a writing or other type portfolio, are assumed to be examples of higher order processes.

The concept of performance assessment is also partially connected to the measurement of higher order thinking (Baker, O'Neil, & Linn, in press). Present formulations of performance assessment involve two major dimensions: the recording of ephemeral behavior—such as electronic trouble-shooting strategy or the conduct of a biology experiment—as well as the rating of resulting products or solutions. It is arguable that records of performance necessarily require higher order thinking. Consider in particular the large proportion of performance assessments in industrial or military contexts requiring the respondent to use specifically practiced operations and procedures in tasks of very narrow boundaries.

One variation on performance assessment that may have a higher order component has renewed currency in public education. The approach, called "seamless" measurement, authentic assessment (Burstall, 1989), or blended assessment (Carlson, 1989), is based on the need for greater validity in terms of both learning and instructional analysis. This approach focuses on the use of complex activities, such as experiments on the absorbency of paper towels (Shavelson, 1989) or the

preparation of analytical papers in history (Baker & Clayton, 1989), to make judgments about achievement. Such tasks may take longer than the usual single class period to accomplish, and may be performed individually, cooperatively, or in a team environment. These activities, even when used for assessment, share many properties with good instructional lessons, including the arousal of curiosity in students and the fact that teachers themselves may be rating the quality of students' performance. In many of these examples, both student processes and products may be judged.

Nevertheless, measures of higher order thinking have significant costs. Among them are the restricted number of tasks that may be sampled in a fixed period of time, the administrative and practical intricacies of reliable and accurate scoring, and the credibility of results for a society used to numbers, stanines, and norms. Alternatively, the benefits of measurement include increased validity and potential for positive curricular and instructional impact. Another important benefit may be the reduction of the salience of ceremonial testing and its attendant costs of targeted test preparation.

Again, the avoidance of a multiple choice format or the presence of a label of open-ended or "authentic" are not sufficient to assure that these measures are actually assessing higher order processes. Details of students' actual instructional histories and the role of test administrators can alter to rote tasks measures that appear to tap the highest reaches of human thought. This reality underscores a major point of this chapter: that any measurement process—higher order thinking in particular—must be understood both in the light of other available information and the intended uses of implementation.

In the next section, the multiple sources for the higher order thinking movement will be discussed. Subsequent sections will consider the policy context for measurement, including a description of the multiple indicator approach. They will also provide detailed examples of both a higher order thinking measure and an example which might inform its understanding in a policy context. Overall, this chapter will consider the multitude of issues relevant to understanding the next wave of achievement measurement.

Impetus for the Measurement of Higher Order Thinking

The potential uses of tests condition their development. In the area of testing in general, and higher order testing in particular, there are at least three major sources for new measures: research, policy, and practice. All three of these have converged on the need of measuring higher order processes. Let us consider each in turn.

Scientific research

The research context impacts measure development in two related ways. First, theoretically driven targets of inquiry—in other words, new constructs—help us reconceptualize our thinking about common processes. Examples of such constructs are: mental model (Norman, 1983; Collins & Gentner, 1984), advance organizer (Ausubel, 1960), "bug" (Brown & Burton, 1978), and metacognition (Meichenbaum & Asarnow, 1978). These constructs are based largely on cognitive science perspectives and provide frameworks to influence the design of human performance measures. Our community has begun to become interested in measuring problem-solving processes, in assessing group performance to capture the social meaning of certain tasks, and in measuring alternative representation modes for student knowledge, all based on scientific research. Still unresolved are the relative roles of process and product in the assessment of higher order thinking, the place of assessment outside or within subject matter domains, and the importance of new conceptions of transfer to assessment.

Clearly, the topics of research are influenced not only by what is theoretically interesting, but by what is socially important as well. Measurement foci are also similarly influenced by social goals. One way to keep score on social importance is to track the availability of research funding. For example, the 1988 *Vincennes* incident in the Persian Gulf, where an Iranian airliner was mistakenly shot down, predicted renewed research attention to measuring task performance of teams under high stress conditions, and such predictions have been verified (see, for example, U.S. Department of Defense (1989), Defense University Research Initiative, Washington, DC, Office). Any choice in R&D goals can only be judged in the light of the trade-offs in support for other goals. A case in point is the de-emphasis on equity issues in the political arena during the Reagan administration. This choice had clear consequences for acceptable assessment foci, consequences which no doubt benefitted more elitist targets such as content-focused studies of higher cognitive performance.

A second, related influence from the domain of research derives from the actual methods used in research itself. When research approaches emphasized behavioral constructs, and quantitative methods with summary estimates of large groups held methodological currency, psychometric research and test development marched in support. Thus, there are clear implications for testing inherent in the shift during the last 15 years to cognitive psychology. With its relegitimating of inferences from small samples (Shulman, 1986), self-report data (Ericsson & Simon, 1984), and other practices drawn from the ethnomethodology side of research (Levine, 1988), we should expect characteristics of tests to develop accordingly (e.g., limited task sampling). Collateral psychometric developments (Bock, 1987), for example, permit this transition by generating credible quantitative estimates for measures based on restrictive task sampling.

Educational policy

The major impetus for identifying higher order thinking as an accountability target grows from policy, not science, however. Recent surveys of educational reform (Pipho, 1988) confirm that tests are used as policy instruments with increasing frequency since the publication of *A Nation at Risk* (National Commission on Excellence in Education, 1983). Tests are seen, correctly, as at least one operational way to communicate standards, and their uses have proliferated, extending beyond requirements for the award of high school diplomas to exit tests from kindergarten, for grade-to-grade promotion, merit diplomas, and teacher recertification. If tests are so important, attending to their focus becomes more critical. Reasons for emphasizing higher order thinking in school programs and assessment have been identified in reports by prestigious and powerful groups. Three ideas recur often in these reports.

The erosion of economic competitiveness. The international trade imbalance, the editorials about the second-tier status of America, and the visible incursion of foreign wealth continue to raise sharp anxieties. These problems have been directly attributed to the failure of schools to prepare an adequately educated workforce in reports by powerful and prestigious groups such as the National Academy of Sciences, National Academy of Engineering, and Institute of Medicine (1984). The comparison group of interest is obviously the Japanese, and the subject matters of concern are mathematics and science. The International Association for the Evaluation of Education Achievement report on U.S. students' performance in mathematics (McKnight et al., 1987), coupled with the emphasis on formal testing in Japanese schools, strengthens for many the argument that testing will provide a way to improve student performance (National Governor's Association, 1986). One might note, with some irony, that the tests used in Japanese schools consist principally of rote items. Thus, it is probably not the test content itself that provides Japan with its educational competitive edge, but rather the relatively permanent consequences of test failure.

The failure of schools to educate disadvantaged students to a level that permits their full and productive participation in our society. Renewed awareness of the problems of disadvantaged — principally black and Hispanic — youth is fueled by media reports of increased violence, gang participation, an alarming rate of teenage pregnancies, and rampant drug traffic and use. National attention is also focused on adult literacy (see, for example, House & Madula, 1987; Kirsch & Jungeblut, 1986; Bain & Herman, 1988; Sticht, 1987; MDC, Inc., 1985).

The inability of the educational system to prepare students for change. Here, the grab bag of societal ills is attributed to failure of individuals and organizations to adapt to change. Issues such as industrial redevelopment, unsuccessful job displacement and retraining, increasing reliance on technology, and illegal immigration are lumped together (see Cohen, 1987; Carnegie Forum on Education and the Economy, Task Force on Teaching as a Profession, 1986; Goodlad, 1984; Sizer, 1984).

The indisputable fact is that these concerns stimulated a rash of specific state and local reform efforts (Bennett, 1988) and that the focus of these reforms substantially has been standards and accountability. How has educational practice reacted?

Educational practice

Let's call educational practice the composite of what local administrators and teachers do to implement policy, to get curriculum enacted, and to teach appropriately in classrooms. Systematic educational models link practice and requirements. These models emphasize the centrality of formulating goal statements, the consequent design and implementation of curricula and instructional practices, and the subsequent creation, administration, and interpretation of measures to assess the attainment of such goals. In real life, these relationships are less rational and orderly. From a period of time when the text materials were the dominant fact of life, we have moved to a period when what is tested is of equal or more importance. At this juncture, practitioner views are predictably divided. Many critics see testing as a pre-emptive, coercive, implicit goal-setting device, whereas other assessment proponents see only good management. The problem used to be understanding whether the major impact of formal testing programs was to *measure* the impact of innovation or to *reform* the educational programs themselves (Baker, 1989). Most policy analysts would probably now adhere to the reform function of tests.

If we believe that assessment now sets goals, we need to examine how assessments and resources are structured to allow teachers to improve their teaching. To do so, tests should be reported diagnostically and componentially connected to curricula, so that teachers can act upon them instructionally. Knowing that a class falls into the lowest quartile is less valuable to a teacher than knowing what domains of content and skills need improvement for given children. A related aspect to the formative evaluation use of tests (Baker, 1974) is the interplay between the availability of texts and other curriculum materials and their relationship to both goals and measured outcomes. This resource issue has been a matter of major focus in some states and districts, sometimes under the label of curriculum alignment. To align a curriculum means that attempts are made to match goals, classroom instructional resources, and tests. When the test is the only element in the alignment set with clear sanctions associated with it, the end result of alignment processes amplifies even more the curricular influence of the test. Depending upon where you sit ideologically and organizationally, this kind of efficacy may be good or evil, as curriculum gets narrowed or focused.

For tests to impact instruction, teachers must be able to teach higher order skills in the first place as well as to understand test results to improve their instructional strategies. Most studies of classroom practice suggest that teachers do not use many higher order teaching practices, although there is evidence that they can be instructed to do so (Pogrow, 1988). Unfortunately, the knowledge base upon which teacher training has drawn is not rich in providing clear strategies and techniques for teachers to use. Furthermore, teachers have limited management options for dealing with increasingly individualized learning requirements, and are again constrained by habit and standard school organization. A related issue is teacher's ability and interest in using test data to revise instruction. Studies of teachers' use of test information show that neither curricula nor teachers' instructional stamina could cope with the level of accuracy, detail, and frequency of information tests could provide (Dorr-Bremme & Herman, 1986). Furthermore, teachers are so used to test information that is irrelevant to the way they perceive teaching that such information is ignored without the pressure of accountability sanctions. Attempts to make higher order thinking assessments appear to be more like classroom activities and less foreign to the teaching environment should help. Furthermore, somewhat down the road, the utility of classroom databases (Herman, 1988) and en-

riched and well-supported computer interventions (Baker, Herman, & Gearhart, 1989) may provide needed support. But computer magic is not yet widespread. Most encouraging may be the opportunity provided by the reprofessionalization and restructuring movements. These reforms, to provide more power to teachers, more than touch on the topic of higher order thinking. Teacher organizations may very well negotiate accountability requirements in the service of focusing on important intellectual goals for children and new teaching behaviors for themselves. The recent activities of the National Board for Professional Teaching Standards (1989) suggests that recertification assessments will strongly focus in this direction. Thus, there is some small chance that the converging interests of science, policy, and practice may actually result in serious reformation of schooling. If other strategies are selected, teachers will have the problem of delivering on policy expectations without changes in their preparation, resources, or commitment.

To sum up the points of this section, major forces have converged, promoting higher order thinking assessment in the research and policy communities in particular, and have raised the stakes for school accountability one more round.

Prospects for Success for Policy-Driven Higher Order Thinking Assessment

What has been recent testing experience? To what extent have prior testing reforms been successful? Answers to these questions may help us to predict if tests of higher order thinking will achieve their intended policy goals. Opinion is mixed. Case studies of testing reforms conducted in five different localities were reported by Ellwein and Glass (1987). They contend that testing programs possess largely symbolic value because the educational system finds a way around standardized testing requirements. Cut scores get lowered and other safety nets are strung up to protect individuals who don't succeed. Shepard, Kreitzer, and Graue (1987), in an analysis of the Texas teacher recertification test, reached similar conclusions, as did Rudner and Baker (in press) when they reviewed statewide teacher testing programs. Others are studying the concomitants of more stringent standards and testing programs. Of interest are the differential effects on minority students, effects which may increase their drop-out rates (Catterall, 1987).

How much testing of higher order thinking is going on? This question is not easily answered because of definitional problems and the aforementioned potential transformation by instruction of higher order skills into memorized procedures. In order to determine the nature and distribution of higher order thinking items in mandated state programs, a study was conducted by Baker, Burstein, Aschbacher, & Keesling (1985) to document the targets of assessment. Of the states that either developed or contracted specially for state testing measures at that time, very few were found to use tests that pushed far beyond tasks of information retrieval or relabeling.

Similarly, very few tests focused beyond the basic skills and assessed knowledge and skills in subject matter areas such as social studies or science. Mathematics assessment dealt largely with arithmetic. We know of specific subsequent changes in testing programs in Texas, California, Connecticut, and New York that are designed to address higher order thinking concerns, and because these states often provide leadership, we will expect to see more tests labeled higher order in the near future.

So higher order testing is on the way. What should we anticipate from investment in the measurement of higher order thinking skills? If one believes Ellwein and Glass (1987), higher order thinking may become little more than a symbolic flag around which to rally. One reason used to justify the development of such measures is that, in time, they will be able to detect the effects of reforms such as tougher curriculum standards (e.g., more time spent in class and more courses in particular subject matters). If testing results are unacceptable (i.e., not enough higher level performance is demonstrated) and the political stakes are high enough, Ellwein and Glass and others believe that the system will find a way to blur outcomes to make them palatable to an expectant public. Such ways include changing pass scores, using less difficult items

nominally to measure a higher order objective, or countenancing forms of explicit practice in instruction, thereby changing higher order skills functionally into retrieval behaviors. Illicit practices, such as falsifying results, also occur. At minimum, we might worry that the term higher order will be misappropriated and used to describe less challenging skills.

Some relief may come from the focus on performance-based activity structures as sources for a new regime of standardized tests. Allied with opportunities provided by the restructuring movement and the leadership of influential states, the future of testing may be brighter than its past.

This entire set of events may be further perturbed by our conditioning to expect reports about achievement in the simplest terms. Fundamental misunderstandings about test norms and about the validity of single-score summaries persist (Cannell, 1988a, 1988b). A mini-industry has developed in educational agencies and private firms to make test scores understandable to parents and the public, leading us to simplify, summarize, and perhaps work against the real goals we have. In no other field with a closely coupled scientific base are major results principally targeted to the least sophisticated consumer. However, the impetus of the educational quality indicators efforts may have something to contribute to the way we conceive, develop, and report higher order thinking performance.

To summarize, a unified voice has emerged from various sectors of the community supporting the measurement of higher order skills as one way to dramatically improve American education and to save us from being third-rate, or worse. But such policy uses make these testing programs vulnerable. Politicians and administrators have pledged to see to it that the schools are accountable and meet the public's expectations. Making such pledges in the dim light of what we know about teaching and learning of higher order thinking is probably both unavoidable and foolhardy. Under conditions where our instructional interventions are likely to be weak, what is left is to massage but the measure?

Assessing Higher Order Thinking: A Research-Based Example

Moving away from general sources and predictions, the next section of this chapter will illustrate an R&D project in the general area of higher order thinking. Our initial task was to develop sophisticated scoring approaches to be used to assess the quality of content in student's writing. As we became immersed in the research process, we broadened our goals to include writing that was in part text based, and began to think of our task as the assessment of deep understanding. We also undertook the project with the expectation that we would extend our work horizontally into other topic areas, vertically to other age ranges, and diagonally to other subject fields.

Task area

We selected the general area of history, then focused on the pre-Civil War era of American history. Our choice was conditioned by the paucity of good available measures in contrast to the universality of the topic in American schools. After some early textbook analyses, we decided that we needed to use primary source materials as part of the stimuli for performance. Our reviews regretfully suggested that the U.S. history texts available in high schools do not treat topics to any degree of depth, thus rendering the search for deep understanding fruitless. We selected speeches as the genre of primary material, since they optimized on authenticity demands and time constraints (Baker, Freeman, & Clayton, 1989).

Thus, we are attempting to explore the construct of deep understanding; our present definition is tentative and relates to the following components and attendant theoretical bases:

(i) Deep understanding requires the activation of thinking processes applied to specific subject matter content (i.e., history topics). These thinking processes depend upon well-known cognitive analyses of learning, including processes such as active construction in the knowledge acquisition process, elaboration, and the integration of meaningful material into existing prior knowledge (see Brown & Campione, 1986, and the comprehensive review by Segal, Chipman, & Glaser, 1985).

(ii) Deep understanding may involve qualitative differences between expert and novice understanding (see Chi & Glaser, 1980). Expert understanding of topics in this area may be premise driven, allusory, and integrative, whereas novice understanding may be more literal and componential (Baker, Freeman, & Clayton, 1989).

(iii) The expression of deep understanding depends upon a sophisticated interplay among three types of knowledge: strategic, procedural, and content (or declarative). Strategic knowledge represents top-level knowledge of the major attributes and relationships of a discipline (i.e., the extent to which interpretation of events in history is context driven and the result of the interrelationships in complex factors such as politics, geography, and economics; see California State Board of Education, 1988.) It also involves the understanding of the role of the historian, argument structures, verification procedures—what is often called process. We use procedural knowledge to describe routines that the student uses to construct answers to our particular format of measures (e.g., how to write an essay). Content knowledge focuses on the elements inside the discipline, the principles, concepts, and facts that provide the manipulable information base. We wish to distinguish further among the contributions to student performance of prior knowledge, instruction, and the text or other text stimuli provided in the measure.

A constructivist view of comprehension suggests that understanding new material is influenced explicitly by prior knowledge and is facilitated by the activation of broad schemata (Rummelhart, 1980; Wittrock, 1981) or the transformation of the language into historical terms, premises, and viewpoints to provide context. Describing patterns of prior knowledge, or mental models, and their effects on comprehension of new material provides another research base against which to assess our progress (see Kieras, 1988; deKleer & Brown, 1983; Brown & VanLehn, 1980; Carpenter, Moser, & Romberg, 1982.)

Our study was initially designed to expand the content quality scoring rubric for essays in subject matter beyond that commonly used in written composition, (Baker, Freeman, & Clayton, 1989). Our efforts at the outset were to attempt to identify the attributes against which student essays (or longer research papers) might be judged. To this end, essays were collected after 11th-grade students read either a Lincoln or Douglas speech. These were scored by two groups of experts. First, teachers trained to use an essay scoring rubric principally focused on "English-teacher" issues (i.e., organization, style) scored the essays; second, a group of history teachers were asked to score only the content knowledge exhibited in the essays and then to isolate the attributes of best and worst essays. Our data showed a remarkable degree of agreement between the two groups of raters, suggesting that matters of expression were swamping the detection of content knowledge. We also had collected think-aloud protocols from teachers, historians, and students who were asked to read the speeches and respond to the essay question. Our analyses of these essays suggested that experts relied heavily on prior knowledge, usually wrote from a premise or specific organizing principle, and used text information for illustration to construct their arguments. Students and some teachers, on the other hand, attended much more specifically to the presented text, sacrificed coherent argument for comprehensiveness of detail, and prepared essays that were less premise driven.

After a series of pilot studies, we have developed a content scoring rubric that incorporates the following elements: (i) use of prior knowledge, (ii) principles, (iii) facts and events, (iv) problem/premise-driven, (v) text information, (vi) interrelationships, and (vii) misconceptions. In addition, an overall impression score for content quality and for essay quality is solicited.

The evolution of our scoring scheme required some revisions in the prompt solicitation. Our most recent study required two full classroom periods to complete our tasks. They included the use of an associational prior knowledge measure, a multiple-choice literal comprehension test based on the text, and task descriptions that ask explicitly for the integration of prior principles and facts with the text material. Our findings have been extremely encouraging, in terms of both construct validity (teacher judgments, standardized test scores, and transcript information) and utility and reliability of the scoring rubric. We are

most interested to see the robustness of the concepts in our scoring rubric across other essay topics (i.e., the Constitution), in extended tasks, such as research papers, and in other subject areas (e.g., reports of laboratory experiments in biology).

Our immediate research plans call for the validation of this measure and its utility at different age levels. We are also refining general specifications for measures of prior knowledge. We also wish to test the generalizability of our scoring dimensions across topics.

Our hope is to develop a scale that will assess multiple aspects of higher order cognitive performance and can be adapted to particular contexts. For instance, the scale might be specially weighted to emphasize the incorporation of new material in an essay, if knowledge acquisition or learning to learn was the major focus. On the other hand, if the measure were used to assess general history knowledge, more emphasis would be placed on organizing premises, prior knowledge, and misconceptions. We are also in the process of exploring the use of hypertext as a way of directly assessing students' representation of content knowledge in essay planning.

Our approach to deep understanding was selected for practical practice as well as theoretical reasons: (i) extended written responses provide opportunity to observe more directly the products of complex thinking processes; (ii) writing in history has strong policy impetus and potential for use; and (iii) Center for Research on Evaluation, Standards and Student Testing (CRESST) staff have experience in the qualitative rating and psychometrics of essays (Baker, Freeman, & Clayton, 1989). Our initial efforts were focused on developing stimulus materials that would meet standards of credibility for historians and history teachers and simulate instructional exposure (i.e., the long passages to be read).

Creating new measures such as our history task is a valuable undertaking, particularly when the scoring scheme for the task results in economies and flexibility. Task validity, mapped on cognitive subject matter and psychometric constructs, is a necessary feature but not sufficient to guarantee utility in policy contexts. Findings from measures need to be presented in a sensible way, ideally so that the decision maker understands not only the level of attainment but also some explanations for reported results. One approach to provide this context has emerged in the guise of educational quality indicator systems.

Understanding Outcomes: Educational Indicators

The metaphor of indicator systems has been adapted from the field of economics (Murnane & Raizen, 1987; Baker & Herman, 1986). Indicator systems combine data in relatively simplified models to improve understanding of the phenomena at hand. An indicator system can help us systematize the kinds of information that we need to conceptualize and to condition our interpretations of test results. Although many writers have attempted to describe indicator systems (see Oakes, 1986, for a fundamental treatment of the topic in education), such systems typically include measures of inputs, processes, and outputs. Inputs consist of the characteristics of students, the socioeconomics of the neighborhood, and the characteristics of teachers. Processes consist of curricula and instructional characteristics, such as courses offered and specific resources available and used. Outputs involve measures of system success, such as our history higher order thinking measure, achievement test scores, dropout rates, admission and persistence in higher education, and measured attitudes. In economics, comparable information about housing starts, money supply, levels of inventory, stock prices, employment figures, etc. provide the data. Our focus will be principally outcome measurement.

Characteristics of Indicator Systems

What are some attributes of indicator systems that make them interesting to apply to the problems in the measurement of higher order thinking?

An indicator may be a composite of many outcome measures. One articulated goal (see Baker, Linn, & Herman, 1985) for the educational achievement measurement agenda in this country is to expand the bandwidth of indicators used to make judgments about educational quality. This means that more than multiple approaches and particular measures can be used to construct an indicator. For instance, instead of focusing on a particular standardized achievement test battery as the major outcome measure for educational programs, a set of indicators would be developed, each the composite of multiple measures. In the area of higher order thinking, especially where there is some disagreement and a weak knowledge base, multiple measures could be combined—for instance, measures of problem identification, structured problem solving, decision making, inductive thinking, planning, and other forms of reasoning (see Arter & Salmon, 1987, for a more complete list). Of course, student products could be assessed as well in portfolios of student best efforts or as benchmark measures of open-ended responses. Research and resource plans will determine how much triangulated or repeated measurement is appropriate. But the existence of multiple measures has benefits beyond increasing validity. When more than a single test is used as an outcome measure, the potential corruptibility of the measure is reduced. The likelihood of illicit test preparation is rendered more difficult. An indicator model also supports the idea of the interactions of individual differences and specific approaches to measurement, and its use emphasizes the multiple perspectives of ideas in this field rather than the monolithic surety communicated by hoary achievement tests.

Indicators are reported in context. Appropriate reporting of any indicator deals concurrently with other available data. Achievement indicators would be reported in the context of other input, process, and outcome indicators. These indicators would be similarly complex, consisting, as appropriate, of measures such as allocations of effort (e.g., changes in enrollment in various courses), student mobility and dropouts, teacher mobility, affective outcomes, per capita support, and class size in the school. The challenge here is to develop indicators of student educational experience to aid in the interpretation of student performance. Although the use of multiple measures is not new and has been a characteristic of many program evaluations, indicator systems are designed to be broadly institutionalized and to be robust across site and program differences.

Indicators derive their meaning over time. Indicators would provide a longitudinal and general estimate of the health of the system — what assessment programs were supposed to do in the first place, rather than posing, as some testing programs are forced to do, as finely honed measures of specific performances of the educational system. The meaning of such an indicator would be derived from changes over time (is it going up or down?) rather than from its absolute value. Also, changes in the make-up of indicators (e.g., the number of integrative, higher order thinking tasks) provide another longitudinal measure of educational quality.

Indicators foster a more participatory educational environment. There should be reduced incentive to tinker with any of the particular measures used to create the composite, since any one measure would be less likely to affect the outcome value. Therefore, specifications and comparable forms for individual measures making up any composite indicator could be widely available. One feature of an ideal system might be the opportunity for local schools, or even classrooms, of teachers and of students to record their particular instructional emphases, both to describe the experiences of students and to help interpret other data. The benefit of indicators is that the summarized, combined information would still be relatively easy to understand for policy and public consumers. Schools would be encouraged to use particular measures as professional aids to the improvement of instruction, in a hypothesis-generation pattern designed to support the intellectual involvement and interest of teachers.

Context for Higher Order Thinking: An Indicators Example

Performance of outcome indicators, such as measures of higher order thinking, are among those that clearly need explanatory contexts. For example, in the recent report of a pilot study of students' performance on open-ended questions in mathematics (California State Department of Education, 1989), fewer than half of the students achieved a satisfactory level of performance and a relatively small percentage produced competent answers. One explanation of the poor performance is that few such types of problems are taught in schools. Similarly, poor student performance can be partially explained by curriculum indicators such as lack of coverage of topics in schools, inadequacies of textbooks, and low student enrollments in relevant courses. An exploration of the development and validation of indicators for use in state systems is underway by a team of UCLA/RAND researchers engaged in developing such indicators in secondary school mathematics and history curricula. One purpose of this project is to develop interim policy indicators to assess the impact of reforms in the area of curriculum standards. Rather than waiting for outcome indicators to show improvement, curriculum indicators are designed to detect the extent to which reform intentions have found their way into enacted curricula. As such, these indicators would provide powerful explanations for subsequent performance levels.

A second goal was to provide a more comprehensive picture of the content classwork and of students' course-taking patterns. At minimum, these measures should respond to policy changes in course requirements.

The complexity of the UCLA/RAND project encompassed issues of how to conceptualize the data, how to validate any new indicator we developed, and how to develop it in an institutional form so that regular data could be acquired without undue burden on the responders. Determining the effects of standards on course taking seems relatively straightforward. For instance, it is common to report data based on the number of students enrolled in courses by title (e.g., Algebra I or American History). A less obvious problem is that the content of a course may be vastly different depending upon who teaches it. Even where attempts are made to standardize course content in a particular district, conventions for what content is included in a given course will differ from site to site. Furthermore, some school districts create courses with even less standardized content (e.g., business math). We hoped to develop an approach that could be used at many sites and to provide a standard framework so that school, district, and state data on course content might be compared.

At the outset, we wished to determine the intent behind the institution of higher coursework requirements. This task was accomplished via a series of focused interviews (Catterall, 1988) with governors' aides, policy makers, and education leaders. In general, these interviews supported our assumptions that more stringent course requirements had been instituted to improve the quality of student performance. McDonnell (1988) reported that policy makers described the purpose of coursework reforms both in general terms — "...kid's potential not being tapped...," — and as more operational goals — "...to raise test scores..." (p. 5).

Our second task was to attempt to find new, comprehensive ways to determine the content of particular courses. Most particularly, we were interested in course content level of difficulty. Our data collection involved five types of data: course enrollment data from school rosters, teacher surveys, student surveys, student transcript analyses, and course materials analyses. McDonnell has reported on the utility of such data, their reliability and validity, and the feasibility of collecting and using such information (McDonnell, Koretz, Catterall, Burstein, & Baker, 1989). To provide a flavor of the work, consider the issue of what it means to have one more course required in a mathematics sequence, with that course intended to increase students' learning of additional mathematics content. However, it is possible that content taught in a Math 1 and Math 2 sequence is simply stretched into a Math 3 course as well. Instruction is slowed down. Does this meet the policy intent? Probably not at first blush, although it is possible that students' overall performance may increase because they have a more extended opportunity to learn a fixed amount of content. To obtain measures of what was actually covered in classes, our project surveyed teachers and students. Another approach involved acquiring samples of student assignments completed by the end of a course and selecting average as well as excellent student work, a post hoc portfolio. Conducting topic analyses of texts and asking teachers to show us how far they covered provided another measure.

Needed analyses are underway to combine findings into composites that either have content validity or can be shown to have construct validity. With curriculum indicators of this sort, we should be able to answer questions about the effects of requirements on students. Do test scores rise because low-performing students have become discouraged with higher standards and have dropped out? Do test scores rise because old content is being learned better over longer periods? Do test scores rise because students are mastering previously unencountered challenging content? Are there short-term dips, as suggested by Koretz (1988), because less-able students are taking harder classes? Do we see a predicted drop in average scores and an increase in the variances of students taking those courses? Developing information of this sort, although a complex, time-consuming task, once institutionalized, can greatly contribute to our understanding of all sorts of student performance.

Summary

This chapter explored the definition and impetus for the measurement attention to higher order thinking skills. Through a detailed description, a model development process was presented. This process relied on strong theory, but was firmly grounded as well in concerns for subject matter validity, feasibility, and credibility. This example led to a discussion of educational indicators as an approach to provide context for results of new measures. A brief example of the development of new curriculum indicators was included to demonstrate the complexity and utility of such efforts.

Acknowledgments

The research supported herein was conducted with partial support from the U.S. Department of Education, Office of Research and Improvement, grant number G00869003.

A version of this chapter was presented at the symposium, entitled "Perspectives and Emerging Approaches for Assessing Higher Order Thinking Skills," at the Annual Meeting of the American Association for the Advancement of Science, San Francisco, January 1989.

The author wishes to thank her colleagues on the project, particularly Marie Freeman, Serena Clayton, Yujing Li, Sheng-Chei Chang, David Neimi, and Pam Aschbacher.

References

Arter, J.A., & Salmon, J.R. (1987). *Assessing higher order thinking skills.* Portland, OR: Northwest Regional Educational Laboratory.

Ausubel, D.P. (1960). The use of advance organizers in the learning and retention of meaningful material. *Journal of Educational Psychology, 51,* 267-272.

Bain, J., & Herman, J.L. (Eds.) (1988). *Making schools work for underachieving minority students: Next steps for research policy and practice.* Los Angeles: UCLA Center for the Study of Evaluation.

Baker, E.L. (1974). Evaluation perspectives and procedures. In W.J. Popham (Ed.), *Evaluation in education.* Berkeley, CA: McCutchan Publishing Corp.

Baker, E.L. (1989). Mandated tests: Educational reform or quality indicator? In B.R. Gifford (Ed.), *Testing and the allocation of opportunity.* Boston: Kluwer Academic Publishers.

Baker, E.L. (1989, March). *What's the use? Standardized tests and educational policy.* Paper presented at the Annual Meeting of the American Educational Research Association, San Francisco.

Baker, E.L., Burstein, L., Aschbacher, P., & Keesling, W. (1985). *Using state test data for national indicators of educational quality: A feasibility study* (Final report to the National Institute of Education). Los Angeles: UCLA Center for the Study of Evaluation.

Baker, E.L., & Clayton, S. (1989, June). *The relationship of test anxiety and measures of deep comprehension in history.* Presentation at the Conference of the Society for Test Anxiety Research, Amsterdam, The Netherlands.

Baker, E.L., Freeman, M., & Clayton, S. (1989). *The measurement of deep understanding* (Report to OERI, Grant G-89-0003). Los Angeles: UCLA Center for the Study of Evaluation.

Baker, E.L., & Herman, J. (1985). *Educational Evaluation: Emergent Needs for Research.* Evaluation Comment, Vol. 7, n.2, p. 1-12.

Baker, E.L., Herman, J., & Gearhart, M. (1989). *The Apple Classroom of Tomorrow: 1988 Evaluation Study* (Report to Apple Computer). Los Angeles: UCLA Center for the Study of Evaluation.

Baker, E.L., Linn, R.L., & Herman, J.L. (1985). *Institutional grant proposal* (Proposal to the National Institute of Education for the Center on Student Testing, Evaluation, and Standards). Los Angeles: UCLA Center for the Study of Evaluation.

Baker, E.L., O'Neil, H.F., & Linn, R.L. (in press). Performance assessment framework. In S.J. Andriole (Ed.), *Advanced technologies for command and control systems engineering.* Fairfax, VA: AFCEA International Press.

Bennett, W.J. (1988). *American education: Making it work.* Washington, DC: U.S. Department of Education.

Bloom, B.S. (1956). *Taxonomy of educational objectives: The classification of education goals. Handbook 1: Cognitive domain.* New York: Longmans, Green & Company.

Bock, R. D. (1987). Comprehensive educational assessment for the states: The Duplex Design. *Evaluation Comment, 1,* 1-15.

Brown, J.S., & Burton, R.R. (1978). Diagnostic models for procedural bugs in basic mathematical skills. *Cognitive Science, 2,* 155-192.

Brown, A.L., & Campione, J.C. (1986). Psychological theory and the study of learning disabilities. *American Psychologist, 14*(10), 1059-1068.

Brown, J.S., & VanLehn, K. (1980). Repair theory: A generative theory of bugs in procedural skills. *Cognitive Science, 4,* 379-426.

Burstall, C. (1989, March). *Authentic assessment.* Paper presented at the California State Department of Education Conference titled "Beyond the Bubble," San Francisco.

Burstein, L. (1989, March). *Looking behind the "average": How are states reporting test results?* Paper presented at the Annual Meeting of the American Educational Research Association, San Francisco.

California State Board of Education. (1988). *History-social science framework for California public schools, kindergarten through grade twelve.* Sacramento: California State Department of Education.

California State Department of Education. (1989). *A question of thinking: A first look at students' performance on open-ended questions in mathematics.* Sacramento: Author.

Cannell, J.J. (1988a). Nationally normed elementary achievement testing in America's public schools: How all 50 states are above the national average. *Educational Measurement: Issues and Practice, 7*(2), 5-9.

Cannell, J.J. (1988b). The Lake Wobegon effect revisited. *Educational Measurement: Issues and Practice, 7*(4), 12-15.

Carlson, D. (1989, June). Planning for authentic assessment. Presentation at the Seminar on Authentic Assessment, Berkeley, CA.

Carnegie Forum on Education and the Economy, Task Force on Teaching as a Profession.

(1986). *A nation prepared: Teachers for the 21st century*. Washington, DC: Author.

Carpenter, T.P., Moser, J.M., & Romberg, T.A. (1982). *Addition and subtraction: A cognitive perspective*. Hillsdale, NJ: Lawrence Erlbaum Associates.

Catterall, J. (1987). *School reform assessment* (Report to OERI, Grant G-86-0003). Los Angeles: UCLA Center for the Study of Evaluation.

Catterall, J. (1988). *School reform assessment: Second quarterly report* (Report to OERI, Grant G-86-0003). Los Angeles: UCLA Center for the Study of Evaluation.

Chi, M.T.H., & Glaser, R. (1980). The measurement of expertise: Analysis of the development of knowledge and skill as a basis for assessing achievement. In E.L. Baker & E.S. Quellmalz (Eds.), *Educational testing and evaluation: Design, analysis, and policy*. Beverly Hills: Sage Publications, Inc.

Cohen, D. (1987). Educational technology, policy and practice. *Educational Evaluation and Policy Analysis, 9*(2), 153-170.

Collins, A., & Gentner, D. (1984). *How people construct mental models* (Technical Report No. 5740). Cambridge, MA: Bolt Beranek and Newman, Inc.

deKleer, J., & Brown, J.S. (1983). Assumptions and ambiguities in mechanistic mental models. In D. Gentner & A.L. Stevens (Eds.), *Mental models*. Hillsdale, NJ: Lawrence Erlbaum Associates.

Dorr-Bremme, D.W., & Herman, J.L. (1986). *Assessing student achievement: A profile of classroom practices* (CSE Monograph Series in Evaluation No. 11). Los Angeles: UCLA Center for the Study of Evaluation.

Ellwein, M.C., & Glass, G. (1987). *Standards of competence: A multi-site case study of school reform* (CSE Technical Report No. 263). Los Angeles: UCLA Center for the Study of Evaluation.

Ennis, R.H. (1987). Testing teachers' competence, including their critical thinking. In *Proceedings of the 43rd Annual Meeting of the Philosophy of Education Society*. Cambridge, MA: Philosophy of Education Society.

Ericsson, K.A., & Simon, H.A. (1984). *Protocol analysis: Verbal reports as data*. Cambridge, MA: The MIT Press.

Gagne, R. (1985). *The conditions of learning (4th edition)*. New York: Holt, Rinehart.

Goodlad, J. (1984). A place called school. New York: McGraw Hill.

Herman, J. (1988). *Multilevel evaluation systems project: Final report* (Report to OERI, Grant G-86-0003). Los Angeles: UCLA Center for the Study of Evaluation.

House, E.A., & Madula, W. (1987). Race, gender, and jobs: Losing ground on employment. Boulder: University of Colorado Laboratory for Policy Studies.

Kieras, D.E. (1988). *What mental models should be taught: Choosing instructional content for complex engineered systems*. Ann Arbor: University of Michigan.

Kirsch, I.S., & Jungeblut, A. (1986). *Profiles of America's young adults*. Princeton, NJ: Educational Testing Service.

Koretz, D. (1989, March). [Comments at Session 30.01, Annual Meeting of the American Educational Research Association, San Francisco.]

Koretz, D. (1988). *The effects of coursework reform: Steps toward a sensitive and valid system of indicators* (Draft). Santa Monica, CA: The RAND Corporation.

Levine, H.G. (1988, April). *Computer-intensive school environments and the reorganization of knowledge and learning: A qualitative assessment of Apple Computer's Classroom of Tomorrow*. Paper presented at the Annual Meeting of the American Educational Research Association, New Orleans.

Linn, R.L., Graue, M.E., & Sanders, N.M. (1989, March). Comparing state and district test results to national norms: Interpretations of scoring "above the national average." Paper presented at the Annual Meeting of the American Educational Research Association, San Francisco.

McDonnell, L.M. (1988). *Coursework policy in five states and its implications for indicator development* (Working paper). Santa Monica, CA: The RAND Corporation.

McDonnell, L.M., Koretz, D., Catterall, J., Burstein, L., & Baker, E.L. (1989, March). *Developing indicators of student coursework*. Paper presented at the Annual Meeting of the American Educational Research Association, San Francisco.

McKnight, C.C., Crosswhite, F.J., Dossey, J.A., Kifer, E., Swafford, J.O., Travers, K.J., & Cooney, T.J. (1987). *The underachieving curriculum: Assessing U.S. school mathematics from an international perspective*. Champaign, IL: Stipes Publishing Company.

MDC, Inc. (1985). The status of excellence in education commissions: Who's looking out for at-risk youth? Chapel Hill, NC: Author.

Meichenbaum, D., & Asarnow, J. (1978). Cognitive-behavior modification and metacognitive development: Implications for the classroom. In P. Kendall & S. Hollen (Eds.), *Cognitive-behavioral interventions: Theory, research , and procedures*. New York: Academic Press.

Murnane, R.J., & Raizen, S.A. (Eds.) (1987). *Im-*

proving indicators of the quality of science and mathematics education in grades K-12. Washington, DC: National Academy Press.

National Academy of Sciences, National Academy of Engineering, Institute of Medicine. (1984). *High school and the changing workplace: The employer's view* (Report of the Panel on Secondary School Education for the Changing Workplace). Washington, DC: National Academy Press.

National Board for Professional Teaching Standards. (Spring 1989). *Certifying and Rewarding Teaching Excellence.* Article in the Carnegie Quarterly, Vol. 34, n.2.

National Commission on Excellence in Education. (1983). *A nation at risk.* Washington, DC: U.S. Government Printing Office.

National Governors' Association. (1986). *Time for results.* Washington, DC: Author.

Norman, D.A. (1983). Some observations on mental models. In D. Gentner & A.L. Stevens (Eds.), *Mental models.* Hillsdale, NJ: Lawrence Erlbaum Associates.

Oakes, J. (1986). Educational indicators: A guide for policymakers (OPE-01). Santa Monica, CA: The RAND Corporation.

Pipho, C. (1988). Academic bankruptcy an account ability tool? Education Week, February 17, 1988.

Pogrow, S. (1988). Teaching thinking to at-risk elementary students. *Education Leadership, 46,* 79-85.

Rudner, L.M., & Baker, E.L. (in press). *Teacher testing: Status and prospects.* Greenwich, CT: JAI Press.

Rummelhart, D.E. (1980). Schemata: The building blocks of cognition. In R.J. Spiro, B.C. Bruce & W.F. Brewer (Eds.), *Theoretical issues in reading comprehension.* Hillsdale, NJ: Lawrence Erlbaum Associates.

Segal, J.W., Chipman, S.F., & Glaser, R. (Eds.) (1985). *Thinking and learning skills: Volume 2: Research and open questions.* Hillsdale, NJ: Lawrence Erlbaum Associates.

Shavelson, R. (1989). *Performance assessment: Technical considerations.* Presentation at the Seminar on Authentic Assessment, Berkeley, CA.

Shepard, L.A. (1989, March). *Inflated test core gains: Is it old norms or teaching the test?* Paper presented at the Annual Meeting of the American Educational Research Association, San Francisco.

Shepard, L.A., Kreitzer, A.E., & Graue, M.E. (1987). *A case study of the Texas Teacher Test: Technical report* (Report to OERI, Grant G-86-0003). Los Angeles: UCLA Center for the Study of Evaluation.

Shulman, L.S. (1986). Paradigms and research programs in the study of teaching: A contemporary perspective. In M.C. Wittrock (Ed.), *Handbook of research on teaching.* New York: Macmillan.

Sizer, T. (1984). *Horace's compromise: The dilemma of the American high school.* Boston: Houghton Mifflin.

Sticht, T. (1987). *Literacy and human resources at work: Investing in the education of adults to improve the educability of children* (HumPRRO Professional Paper 2-83). Alexandria, VA: Human Resources Research Organization.

Wittrock, M.C. (1981). Reading comprehension. In F.J. Pirozzolo & M.C. Wittrock (Eds.), *Brain, cognition, and education. New York:* Academic Press.

2

A New World View of Assessment in Mathematics

THOMAS A. ROMBERG, E. ANNE ZARINNIA, and KEVIN F. COLLIS

In this chapter, we consider the consequences of the emerging world view on assessment of students' knowledge of mathematics and their ability to use that knowledge both creatively and routinely in solving the variety of problems encountered in the course of life. The crucial point is that the world is changing so rapidly that, unless those involved in mathematics education adopt a proactive view and develop a new assessment model for the twenty-first century, the mathematical understanding of children will continue to be inadequate into the future.

An implicit premise is that assessment, which has usually involved some testing procedure, has an impact on curriculum and instruction. It is fact that the content and procedural emphasis of assessment has a direct impact both on what is taught and how it is taught (Romberg, Zarinnia, & Williams, 1989).

At present, the nature, forms, purposes, and design of assessment are dominated by the prevailing, old world view helping to perpetuate it. The current structure of teaching and learning was a product of its times. It grew out of the machine-age thinking of the industrial revolution of the past century. The intellectual contents of the machine age rested on three fundamental ideas. The first was *reductionism*. To deal with anything, you had to take it apart until you identified its simplest parts. The second fundamental idea was *analysis*. If you have something you want to explain or a problem that you want to solve, you start by taking it apart. You break it into its components, you get down to simple components, then you build up again. The third basic idea was *mechanism*. All phenomena in the world can be explained by stating cause-and-effect relationships. The primary effort of science was to break the world up into parts that could be studied to determine cause-and-effect relationships. The world was conceived of as a machine operating in accordance with unchanging laws.

These ideas gave rise to the industrial revolution. In this world, work was conceived of in physical terms, and mechanization was about the use of machines to perform physical work. Man was supplemented by machines. Man-machine systems were developed for doing physical work in such a way as to facilitate mechanization.

This whole process is clearly reflected in current school mathematics. Mathematics was segmented into subjects and topics, eventually down to its smallest parts — behavioral objectives. At this point, a hierarchy was created to show how these components were related to produce over time a finished product. Next, the steps by which one traveled that hierarchy were mechanized via textbooks, worksheets, and tests. Furthermore, teaching was dehumanized to the point that the teacher had little to do but manage the production line. Businesses, industry, and, in particular, schools have been conceived and modified on the basis of this mechanical view of the world. Today, assessment procedures are dominated by the forms and functions of this old world view. However, progress toward a new curriculum will be impeded unless the current process of assessment is changed. Consequently, it is essential to lay bare the ways in which contemporary assessment procedures are redolent of the old world view and to point to alternatives.

To develop this argument, we begin by exploring the need for an approach to assessment which is consonant with the reform movement in mathematics education (e.g., the NCTM *Standards*, 1989). Next, we examine the current framework for the profile assessment of knowledge — content by behavior matrices. This is followed by our argument as to why this approach is no longer appropriate in the light of the new world view. In turn, we then summarize new directions.

Need for a New Approach to Assessment

Sometimes, educational reform is directed toward making schooling more efficient. Under those conditions, expected outcomes do not change, and assessment procedures may remain the same if they reflect those expectations. However, if expectations change, new assessment procedures must be developed. This can only be done by comparing and contrasting the old and new expectations, using the assessment tools designed for both, discarding those no longer appropriate, and developing new procedures when needed. Today, schools should be planning to change the emphasis from drill on basic mathematical concepts and skills to explorations that teach students to solve problems, to communicate, to reason, to interpret, to refine their ideas, and to apply them in creative ways.

The need for new assessment procedures which reflect these changes to a new world view is based on five assumptions.

Assumption 1: We are now in a new age — the information age — which will significantly alter the character of American schooling. Some of the attributes of the shift from an industrial society to an information society are (i) it is an economic reality, not merely an intellectual abstraction; (ii) the pace of change will be accelerated by continued innovation in communications and computer technology; (iii) new technologies will be applied to old industrial tasks at first, but will then generate new processes and products; and (iv) basic communication skills are more important than ever before, necessitating a literacy-intensive society.

In 1987, Zarinnia and Romberg argued:

The most important single attribute of the Information Age economy is that it represents a profound switch from physical energy to brain power as its driving force, and from concrete products to abstractions as its primary products. Instead of training all but a few citizens so that they will be able to function smoothly in the mechanical systems of factories, adults must be able to think... This is significantly different from the concept of an intellectual elite having responsibility for innovation while workers take care of production. (pp. 23–24)

In fact, since, on average, most workers will change jobs four to five times, they can no longer assume that they can acquire initially the mathematical skills needed throughout their years in the workplace. A flexible workforce capable of lifelong learning is now required.

Assumption 2: Higher order thinking skills must be the focus of instruction in mathematics. All learning involves thinking, but in the past, most instruction focused on learning to name concepts and follow specific procedures. Now, the emphasis for all students must shift to communication and reasoning skills. Although these skills resist precise definitions, they are now popularly called "higher order" thinking skills. Lauren Resnick (1987) listed some of their features, many of which are in stark contrast to current mathematics criteria (shown in parentheses).

(i) Higher order thinking is nonalgorithmic. That is, the path of action is not fully specified in advance. (still largely algorithmic)

(ii) Higher order thinking tends to be complex. The total path is not "visible," from any single vantage point. (standard examples with visible paths)

(iii) Higher order thinking often yields multiple solutions, each with costs and benefits, rather than unique solutions. (single, unique solutions)

(iv) Higher order thinking involves nuanced judgment and interpretation. (neither judgment nor interpretation expected)

(v) Higher order thinking involves the application of multiple criteria, which sometimes conflict with one another. (simplified to single criteria that are well defined in content)

(vi) Higher order thinking often involves uncertainty. Not everything that bears on the task at hand is known. (certain — all information required is given)

(vii) Higher order thinking involves self-regulation of the thinking process. We do not recognize higher order thinking in an individual when someone else calls the plays at every step. (external regulation)

(viii) Higher order thinking involves imposing meaning, finding structure in apparent disorder. (meaning is given or assumed)

(ix) Higher order thinking is effortful. There is considerable mental work involved in the kinds of elaborations and judgments required. (work which usually involves standard exercises is simplified so that little effort is needed) (pp. 2–3)

Given these contrasts, thinking skills must be the focus of instruction in mathematics in the near future, and assessment procedures need to be developed which portray not only the number of correct answers students can produce, but also the thinking that produced those answers. These features of higher order thinking point to a historical fact that helps to resolve the question of what is new in this shift in emphasis. American schools, like public schools in other industrialized countries, have inherited two quite distinct educational traditions — one concerned with elite education, the other concerned with mass education. These traditions conceived of schooling differently, had different clienteles, and held different goals for their students. Mass education involved mastery of what were considered the basic skills (the 3 R's) needed to participate in a democratic, industrial society. Elite education involved preparation for economic and social leadership. This included the acquisi-

tion of higher order thinking skills with a liberal arts emphasis. Only in the last 60 years or so have the two traditions merged, at least to the extent that most students now attend comprehensive secondary schools in which several educational programs and student groups coexist. Yet, a case can be made that the continuing and as yet unresolved tension between the goals and methods of mass and elite education produces our current concern regarding the teaching of higher order skills.

Assumption 3: Higher order skills need not be learned after other skills. Again, Resnick (1987) has stated:

The most important single message of modern research on the nature of thinking is that the kinds of activities traditionally associated with thinking are not limited to advanced levels of development. Instead, these activities are an intimate part of even elementary levels of reading, mathematics, and other branches of learning — when learning is proceeding well. In fact, the term "higher order" skills is probably itself fundamentally misleading, for it suggests another set of skills, presumably called "lower order," needs to come first. (p. 8)

Assumption 4: All students can learn higher order thinking skills. If mathematics is approached as something one does — "gathers, discovers, or creates knowledge in the course of some activity having a purpose" (NCTM, 1989, p. 7) — then all students should be able to explore, conjecture, and reason.

Assumption 5: The contemporary approach to achievement testing is a conservative inhibitor to needed reform. Les McLean (1982) has stated that "achievement tests as we have known them are obsolete and teachers should discontinue their use as soon as possible" (p. 1). Peter Hilton (1981) argued the issue even more strongly when he states that current tests

... force students to answer artificial questions under artificial circumstances; they impose severe and artificial time constraints; they encourage the false view that mathematics can be separated out into tiny water-tight compartments; they teach the perverted doctrine that mathematical problems have a single right answer and that all other answers

are equally wrong; they fail completely to take account of mathematical process, concentrating exclusively on the "answer." (p. 79)

Clearly, a most important challenge facing the movement for increasing higher order skill learning in the schools is the development of appropriate evaluation strategies. Pressures for use of current tests, especially at the elementary and secondary school levels, make survival difficult for curriculum reforms that do not produce test score gains. Most current tests favor students who have acquired much factual knowledge and do little to assess either the coherence and utility of that knowledge or students' ability to use it to solve problems.

To conclude this first section, four points should be emphasized. First, the educational system as a whole—and the teaching and learning of mathematics in particular—need to be changed. Current reform efforts must encompass more than simple reactions to current weaknesses. To remedy weaknesses, we cannot return to past methods of curriculum development, teacher training, and pupil assessment that were used in the past. Unless these, too, are changed, the same difficulties of sterile lessons, further de-skilling of teachers, and so on will have been created.

Second, information on student performance is important for educational decision making and monitoring the effects of change. There is no question that valid data could and should influence decisions. Clearly, if the content of courses and the methods of instruction change, the monitoring of student achievement is necessary if the effects of these changes are to be determined.

Third, current testing procedures are not likely to provide valid information for decisions about the current reform movement. Current tests reflect the ideas and technology of a different era and world view. They cannot assess how students think or reflect on tasks, nor can they measure interrelationships of ideas.

Finally, work needs to be started on new assessment procedures. Only by having new assessment tools can we provide educators with appropriate information about how students are performing with respect to the goals of the reform movement.

Old World Framework

Achievement tests are often comprised of items that reflect the combination of two classifications. One is related to the content of the items, the other to the behavioral outcomes response. Classification is a fundamental intellectual activity that underlies most practical and theoretical activities. Classification of objects in a domain starts with the broadest, most inclusive categories and progressively subdivides. At any given level in the resultant subsets of subject matter, categories are theoretically mutually exclusive. Each subset may be subdivided according to some principle of internal coherence until a set containing only one object is reached. Equally, subsets may be recombined to reform the initial set. This process of ordered set division, the larger set being an aggregation of its own subsets, is the organizing principle of all hierarchies. It is a method of analysis that has been used on everything from land distribution to library collections. In the process of outlining the work of students and teachers, the principles of classification have also been applied to both the organization and the sequencing of the content to be taught and learned (e.g., Thorndike, 1904; Tyler, 1931) and, with the behavioral objective movement, to the behaviors exhibiting orders of understanding (e.g., Bloom, 1956).

Content

Classification of mathematical content typically depends on the identification of mathematical objects and their attributes. At the broadest level, categories of mathematical content are a convenient way of dividing knowledge into such large chunks as semester courses, textbooks, and major examinations. At an intermediate level, the categories may be used to organize chapters in the textbook or weeks in the course. At an even more specific level, small independent categories of content are the organizing principle for parts of the daily lesson plan, a unit in the text, or a homework assignment. Such categories are advantageous in that they break work into manageable chunks and restrict teaching to the presentation of a clearly defined segment of the content. By structuring content in this manner, it is possible to ensure comprehensive coverage of the subject, whether in teaching, testing, or learning.

Unfortunately, the classifications on which the sequencing of instruction and consequent assessment have been based are largely spurious, a means toward the linear ordering of work. Note that most instructional sequences have been constructed for purely practical reasons, and are not true hierarchies. Often strands, and subjects within strands, are specified, but no conceptual or psychological dependence is apparent or assumed. If a strict, conceptually based partial ordering of the segments could be found, a content hierarchy might be constructed. However, if the structure of instruction and assessment is to have a positive influence, mathematical content must be arranged, where appropriate, in structures based on the interdependence of skills and concepts.

Behaviors

The power of classification as a logical organizer also appealed to college examiners looking for a theoretical framework to facilitate communication. Thus, at the 1948 American Psychological Association (APA) convention in Boston, after considerable discussion, there was agreement that such a theoretical framework might best be obtained by classifying the objectives of the educational process. The resultant taxonomy and nomenclature (known universally as Bloom's *Taxonomy*) were intended to improve communication in the educational community because the objectives provided the basis around which curricula and tests could be built (Bloom, 1956). The proposal rested on the premise that educational objectives stated in behavioral terms have their counterparts in the behavior of individuals, which can be observed, described, and, therefore, classified. However, fear was expressed that

... It [the Taxonomy] might lead to fragmentation and atomization of educational purposes such that the parts and pieces finally placed into the classification might be very different from the more complete objective with which one started. (Bloom, 1956, pp. 5–6)

Nevertheless, Bloom felt that the structure of the hierarchy would enable users to under-

stand clearly the place of objectives in relation to each other. It ignored what is now known to be the unpredictability of human cognitive functioning. Consequently, the taxonomy was formally presented at the Chicago meeting of the APA in 1951 and subsequently published in 1956. (Bloom, 1956). The taxonomy of cognitive behaviors complemented the classification of content as an organizing tool.

Content-by-behavior matrix approach

The content-by-behavior matrix was an even more powerful organizing structure. It permits a rapid overview of the entire structure and relative emphases on particular parts. Consequently, despite modification of the specifics on each axis, the matrix approach has persisted. It was integral to the model of mathematics achievement in the National Longitudinal Study of Mathematical Abilities (NLSMA) (Romberg & Wilson, 1969, pp. 29–44) and the National Assessments of Educational Progress (NAEP, 1981).

NLSMA, for example, originally considered "an eleven-by-seven content-behavior matrix" (Romberg & Wilson, 1969, p. 35). However, content was combined and reduced to three categories: number systems, geometry, and algebra. The behavioral axis was consolidated from seven categories to four: computation, comprehension, application, and analysis. In the second International Education Assessment (IEA), Weinzweig and Wilson (1977) recommended a matrix identical to the NLSMA on the behavioral axis, but subdivided into nine categories on the content axis. By comparison, the second NAEP, using a content-by-process matrix, divided the behavioral axis (process) into knowledge, skill, understanding, and application. The third assessment expanded application to include problem solving and added an attitude category (NAEP, 1981). Persistence of the matrix as a tool for organizing activity is important and reflects (i) its power as an organizing tool, (ii) its visual facility, and (iii) the strong continuity among assessment projects created by relying on those with the most relevant experience in the field when planning the next project.

In summary, the theoretical model implicit in major evaluations of mathematical education has been based on a matrix of taxonomies of content and behaviors. The convenience and power of the model is reflected in its persistent use in the face of changing circumstances. However, its inadequacies have begun to outweigh its conveniences.

Discontent

Noting the failure of the mathematics reform efforts of the 1960s and early 1970s, Westbury (1980) argued that change involves the abandonment of some practices and the adoption of others. The deep structures of formal and informal institutional apparatus, procedures, forms, and rituals tend to preserve the status quo, frustrating efforts at curricular reform. However, just as students have difficulty in learning because they fail to modify old conceptions, so also do ingrained theoretical structures carry an intellectual baggage that impedes change.

A new cohesion among the goals of education, its practices, and methods of assessment that would promote educational change, rather than stifle it, therefore depends on the divestiture of old styles of thinking. For that to happen, there must be a recognition of the ways in which the concepts of the old world view dominate the deep structure of present evaluation. Despite long-standing and growing concern, the values and forces that dominated mathematics education for the past century are embedded in the theoretical structures of prevailing methods of assessment.

Behaviorism

Behaviorism reflected an application of the engineering approach of scientific management to the problems of education, focusing on managing environmental factors to achieve a defined outcome and ignoring the internal cognitive mechanisms. Scientific management rested on three basic principles: specialization of work through the simplification of individual tasks, predetermined rules for coordinating the tasks, and detailed monitoring of performance (Reich, 1983). These microprinciples pervaded American education with the same thoroughness with which they were applied in the economy. They dominated the breakdown of knowledge, the roles of teacher and students, instructional and administrative

processes, the building-block approach of Carnegie units, the content and structure of textbooks, belief in the textbook as an effective tool for transmitting content, the structure of university education, and monitoring and evaluation. Out of this content emerged the notion of progress through the mastery of simple steps, the development of learning hierarchies, explicit directions, daily lesson plans, frequent quizzes, objective testing of the smallest steps, and scope-and-sequence curricula.

Unfortunately, these are only the more obvious aspects. One consequence of such meticulous planning is that it renders the unplanned unlikely. A second is that a system designed to eliminate human error and the element of risk also eliminates innovation. A third is that, like factory work, it is dull, uninspiring, and unmemorable except for the boredom associated with it. Personal involvement and the mnemonics of the unexpected are nonexistent.

Bloom's taxonomy of educational objectives epitomizes the domination of American education by scientific management, for it completed the process by which not only the content of learning, but also the proxies for its intelligent application, were classified, organized in a linear sequence, and, by definition, broken into a hierarchy of mutually exclusive cells. The consequences in the classroom were far reaching. Scope and sequence charts prescribed which parts of a subject were to be covered in what order; each cellular part of each subject was put into a matrix (e.g., Romberg & Kilpatrick, 1969, p. 285); behaviors suggesting desirable intellectual activity were also sequenced. However, given the multiplicity of subject cells to be covered, the easiest way to finish the prescribed course of study was to simply cover content without worrying too much about thought. Furthermore, matrices are difficult to construct effectively on paper in more than two dimensions. Consequently, few scope-and-sequence charts addressed in a coherent manner both levels of thinking and specific aspects of content within an overall discipline. Thus, one focus of concern in documents addressing the quality of education has been the failure of students to attain "higher order intellectual skills" (National Commission on Excellence in Education, 1983, p. 9). Continuation of this pattern will be catastrophic because it suggests that those responsible for evaluation have failed to recognize the power of the deep structures to constrain curriculum development (Westbury, 1980) through the implicit goals suggested by the form and content of evaluation.

Attacking behaviorism as the bane of school mathematics, Eisenberg (1975) criticized the dubious merit of a task-analysis, engineering approach to curricula because it essentially equates training with education, missing the heart and essence of mathematics. Expressing concern over the validity of learning hierarchies, he argued for a reevaluation of the objectives of school mathematics. The goal of school mathematics is to teach students to think, to make them comfortable with problem solving, to help them question and formulate hypotheses, investigate, and simply tinker with mathematics. In other words, the focus is turned inward to cognitive mechanisms.

The persistence of Bloom's intellectual model is also reflected in the continued use of associated nomenclature. Use of the term higher order thinking, for example, directly expresses reliance on Bloom's taxonomy for the theoretical model. This is of particular concern because of its associated intellectual baggage; it implies that lower order thinking precedes higher thinking processes. However, activities associated with higher order thinking are not limited to advanced levels of development. Failure to stress higher order features of thinking because of the belief that a lower order must be attended to first is a source of major learning difficulties. In reading, for example, cognitive science has suggested that "processes traditionally reserved for advanced students...might be taught to all...especially those who learn with difficulty" (Resnick, 1987). This approach is subliminally impeded by continued reliance on nomenclatures and models of assessment that have Bloom's taxonomy as their underlying construct.

Content

By definition, classifications of knowledge, whether for the purpose of organizing the curriculum or for monitoring curricula, make an implicit statement of theory. Definitions of curricula that focus on knowledge divided into subjects for study—such as mathematics is divided into arithmetic, algebra, and geometry—have the immediate impact of stating that (i) knowledge can be separated into clearly defined, independent, self-sustaining parts; (ii) such an approach is important, more important than any other approaches that might follow; (iii) there is a logical sequence of development in which each part builds on a preceding foundation; (iv) it is important to know about the divisions of knowledge enumerated; and (v) if knowledge is learned in this manner, students would be able to use and apply their mathematical knowledge as needed.

Such implicit assumptions may be unwarranted if, for example, knowledge is regarded as unitary and emphasis is on knowing rather than knowing about. The behaviorist approach is also unsuitable if there is genuine concern with application and problem solving. Stated simply, purpose should suggest form, and form implies purpose; incoherence may be inferred from anything less.

Disagreement over the precise structure and arrangement of content in a grid is only part of the problem. Westbury (1980) pinpointed a more fundamental concern: the difference that exists between the intellectual structure of a discipline and its institutional structure in schools, where it is an administrative framework for tasks. The consequence is that administrative stability impedes intellectual change. For similar reasons, Romberg (1985) described mathematics in schools as a stereotyped, static discipline in which the fragments have become ends in themselves. A similar response to the impact of scientific management and behaviorism on mathematics as a school subject is Scheffler's (1975) denunciation of the traditional, mechanistic approach to basic skills and concepts:

The oversimplified educational concept of a "subject" merges with the false public image of mathematics to form quite a misleading conception for the purposes of education: Since it is a subject, runs the myth, it must be homogeneous, and in what way homogeneous? Exact, mechanical, numerical, and precise — yielding for every question a decisive and unique answer in accordance with an effective routine. It is no wonder that this conception isolates mathematics from other subjects, since what is here described is not so much a form of thinking as a substitute for thinking. The process of calculation or computation only involves the deployment of a set routine with no room for ingenuity or flair, no place for guesswork or surprise, no chance for discovery, no need for the human being, in fact. (p. 184)

Multiple-choice format and item independence

The single, most severe criticism of objective test questions that are designed to assess a specific item of content at a specific level of behavior is that they trivialize learning and knowledge (Berlak, 1985). For several reasons, this is almost inherent in such questions. First, they are constructed to test a single, specific objective, clearly defined in the matrix. Thus, elements in the multiple-choice format are designed so that the candidate can pick an answer which is sufficiently specific to unequivocally demonstrate the sought behavior. This tends to eliminate any synthesis between content and behavior. Second, the very nature of objective tests, which ask the user to choose among alternatives, eliminates creativity in answering. Even the intent militates against creativity because it is micro-analytical rather than synthetic or creative.

Frederiksen (1984) observed that a multiple-choice format does not measure the same cognitive skills as a free-response form, and that "efficient tests tend to drive out less efficient tests, leaving many important abilities untested — and untaught" (p. 201). One example of a desirable outcome untested and untaught is the ability to cope with ill-structured problems, which are not found on standardized achievement tests.

A less obvious impact was observed by the Assessment of Performance Unit (Cambridge Institute of Education, 1985). The multiple-choice format is an interventional mode of questioning which appears to offer a greater chance for success in situations where the student is unfamiliar with the material. However, in other situations, students benefitted from the opportunity to think, achieving greater success with the free-form response.

Another aspect of most objective tests is

that, even though some questions may be designed to test lower level thinking and others are designed to evaluate higher thought processes, they are usually tested independently of each other, allowing little notion of a student's approach to a given problem.

In addition to their direct effects, such tests exert powerful indirect effects on both the style of teaching and the style of learning. When one studies for an essay exam, one progressively surveys and synthesizes, putting the parts together and developing a mental model of the structure of the subject. One also develops points of view and arguments to advance and support, for those are the expectations. By contrast, in studying for an objective, multiple-choice test, one learns to cover the parts and to make fine distinctions between alternative ways of stating the same thing in order to distinguish a "right" answer from a "wrong" one, the implication being that there is a single right answer. In other words, the one form requires that students create their own models of mathematics, the other reinforces the view of mathematics as ground to be covered.

Summary

Unfortunately, it is incredibly difficult to shrug off old habits. For example, the architecture of current evaluations — the two-dimensional, content-by-behavior matrix — is a seductively convenient model for organizing information visually and for reducing the inherent complexity of relationships, which we know give a false sense of simplicity. Occasionally, a three-dimensional version expands the possibilities (e.g., Carpenter, Coburn, Reys, & Wilson, 1978; Foxman, Cresswell, & Badger, 1981), but increases the conceptual load and so is used less. The intellectual consequences of using a two-dimensional matrix deserve thought. Such a matrix encourages a tendency to tacitly view successive cells in a row or column as entities having a sequential and linear relationship to each other. It also causes visual separation of nonadjacent cells, subliminally interrupting perception of relationships between them. If such relationships do exist, the visual patterns of the matrix have a powerful, often mnemonic, impact. If not, the framework is not inert; it suggests relationships that are simple, numerically restricted, and linear. Persistence of the matrix form is likely to continue as long as information is presented on paper. However, the potential of electronic data bases and computer-based modeling suggests that multiple viewpoints are more revealing and less constraining.

New Directions

In any model, cohesion comes from purpose. The tight cohesion of the content-by-behavior framework originated in the intent to assess and, implicitly, sort students according to their knowledge about mathematics and, secondarily, by their ability to think. It is a quantitative, linear model involving content, process, and people. The content-by-behavior matrix of evaluation does not question the purpose for teaching mathematics — it reflects purpose. It also derives from congruence between purpose and intellectual tools. The purpose was to sort people into linear rankings of extensionally definite sets, which is precisely what a matrix does.

Evaluation and assessment in a stable paradigm may take for granted the purpose for teaching and the philosophical foundations of the subject under evaluation, but in a period of major societal change, such nonchalance is unwise. The new world view specifically rejects the consequences of old cohesion, of which the content-by-behavior matrix is a microcosm. Because the stated purpose is to cooperate in the creation of mathematical power for each student, that concept should become the cohesive force of any new theoretical model. It is, furthermore, a qualitative, rather than quantitative, concept.

The new world view is, above all, integrative; it sees everything as part of a larger whole, with each part sharing reciprocal relationships with other parts. It seeks a rational balance between education and training, between cooperation and individual effort, between the development of intelligence and its measurement, between the integration of intuitive and analytical thinking and an exclusive stress on the analytical, and

between constant learning for the purpose of innovation and adaptability as opposed to one-time schooling for life. The new world view stresses the acquisition of understanding by all, including the traditionally underprivileged, to the highest extent of their capability, rather than the selection and promotion of an elite. It is a philosophy that simultaneously stresses erudition and common sense, integration through application, and innovation through creativity. Most importantly, it stresses the creation of knowledge. It is as tightly coherent as the old world view. To espouse the intent but retain the old model of assessment is to lose the integrity of the old without gaining that of the new.

To recap, the process of assessment affects the educational process that it is designed to evaluate. The power of the old model derived in large part from its congruence with the underlying and coherent philosophies of science and society. Cohesion is a matter of purpose. Logically, to be a powerful tool for intervention, any new model should be as closely congruent with the purposes, philosophy, and methods of the new world view as the matrix model was with the old. At present, we are not in a position to describe a new model. However, we can outline some alternate assessment procedures which appear to be consistent with the new world view.

Alternative assessment procedures

We believe that instruments for assessment should embody the following commonalities: (i) all knowledge is rooted in experience; (ii) knowledge entails the structural modelling of perceived regularities and the reconciling of irregularities; (iii) cohesion of structure is integral and derived from purpose; (iv) quality is determined by predictive power; (v) disequilibrium is essential to the process; (vi) knowledge is both individual and communal.

Simply stated, there is a need for tools that document the production of knowledge and not merely the proxies that contribute to the process, such as time spent learning or the quality of the teaching staff. A sufficiently detailed view of the process is essential in order to have some idea of how to construct policies for intervention. However, if there is any lesson to be learned from the old paradigm, it is that parts of the process cannot be analyzed in isolation, and then aggregated, with the result regarded as an adequate indicator.

Several approaches offer some promise. Two are discussed here, one that was developed in this country and one in Britain. The first approach consists of the two types of items based on the SOLO taxonomy (Biggs & Collis, 1982). Of these two types, one has an open response format where the student is given a problem to solve, and the assessor first maps or tabulates the selection and use of the data given and the processes and concepts used by following the individual's reasoning as it is set down. This can then be followed up, and the map or table adjusted during an individual interview. The other type based on the SOLO taxonomy, is the "superitem" (Collis, Romberg, & Jurdak, 1986), a closed form of the open response item in which the student responds to a sequence of questions. A correct response to a question indicates a certain use of the data and the availability of certain concepts and techniques.

A second is the constellation of innovative approaches being tried in Great Britain. These incorporate pencil-and-paper testing, practical testing, diagnostic interviewing for the identification of strategies and errors in problem solving, and the effort to develop graduated assessment in mathematics. Both approaches show promise for offering information which will not only facilitate judgments about the student's current level of mathematical thinking, but will also provide information needed for diagnostic teaching and are consonant with our new world view paradigm.

SOLO taxonomy approach

An analysis of several theories of cognitive development put forward during the last decade (Biggs & Collis, 1982 and in press; Case, 1985; Demetriou & Elfklides, 1985; Fischer, 1980; Halford, 1982; Mounoud, 1985) would seem to show consensus on certain key issues of relevance here. Although there are important differences between the theories, the weight of evidence appears to support the original SOLO analysis (Biggs & Collis, 1982), which suggested that there were two phenomena involved in determining the level of an individual's response to an environmental

cue. The first they have called the mode of functioning (Biggs & Collis, 1989). It is characterized by the degree of abstraction that is utilized by the learner in handling the elements of the task involved and is closely related to the existing notion of Piaget's stages of development. The second, termed the Structure of the Observed Learning Outcome (SOLO), is defined by the level attained in the "cycle of learning" which is common to each mode. As mentioned earlier, modes are levels of abstraction; they progress from concrete actions to abstract concepts and principles and correspond in large part to Piaget's stages of cognitive development — sensorimotor (birth to 2 years), intuitive/ikonic (2 to 6 years), concrete symbolic (7 to 15 years), and formal (16 + years) — in which each stage has its idiosyncratic mode of functioning and, as far as intellectual functioning is concerned, its own set of developmental tasks. It is important to note that the modes do not successively replace each other, as is suggested in classical stage theory, but as each develops, it is added to its predecessor (which itself continues to develop), and, thus, the modal repertoire of the mature adult is considerably greater than that of the young child.

SOLO, on the other hand, is concerned with describing the structure of any given response as a phenomenon in its own right — that is, without necessarily representing a particular stage of intellectual development. As the learning cycle within each stage develops, the structure of the response becomes increasingly complex. Prestructural responses represent no use of relevant aspects of the mode in question; unistructural responses represent the use of only one relevant aspect of the mode; multistructural responses represent several disjoint aspects, usually in a sequence; relational responses involve several aspects related into an integrated whole; and an extended abstract response takes the whole process into a new mode of functioning. These notions may be best summarized by considering the diagram in Figure 2.1.

Ages at which modes typically first appear are indicated on the abscissa; the modes themselves accrue as indicated on the ordinate. Once each has developed, it remains as a potential medium for learning throughout life. The learning cycle progresses from

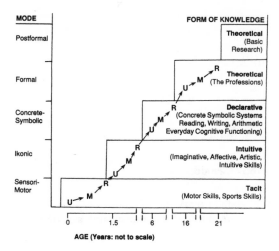

Figure 2.1. Modes, learning cycles, and forms of knowledge (adapted from Biggs & Collis, in press).

unistructural (U) through multistructural (M) to relational (R) within each mode; the extension from relational to extended abstract, becoming unistructural in the next mode. Within the rows, an indication of the nature of the contents learned in each mode is shown, and the form of knowledge associated with each mode.

Perhaps the most important feature of this model is the marriage between the cyclical nature of learning and the hierarchical nature of cognitive development. Each level of functioning within a cycle has its own integrity, idiosyncratic selection, and use of data; yet, each provides the building blocks for the next higher level. The movement from relational to extended abstract within a cycle marks the transition to a new mode of functioning, a new stage of development. The higher mode then proceeds through a similar structural reorganization until, given the individual's continued development, it eventually provides the necessary conditions prerequisite for the development of the next mode and so on.

Thus, there are three factors involved in the process of intellectual functioning: (i) cognitive development, as represented by movement up the modes; (ii) the continuing availability and development of each mode once developed; and (iii) the increasing complexity of the structural organization within a mode as it is developing. All of these are of considerable importance when we come to as-

sessing student responses. However, it is the last of these at which the SOLO taxonomy is basically targeted.

As originally developed, the SOLO taxonomy used an open response format in which student responses were examined for structural organization by an assessor. A later development (Collis & Romberg, 1981) enabled the technique to be used in a closed format. Let us look at some examples of these two formats.

SOLO taxonomy: Open format

In this form, the student is either given information and asked a question requiring a response, or given a task that requires the student to draw on his/her long-term memory store for suitable data to complete the task. The open technique has always presented some difficulty in categorizing the responses in mathematics since the student's response does not necessarily indicate how the material was manipulated to obtain the result.

However, some promising techniques are in the process of development to assist in this process (Collis & Watson, 1989; Chick, 1988; Chick, Watson & Collis, 1988; Siemon, 1988). One involves a mapping procedure (see the Chick, Collis, and Watson studies just mentioned) that sets out to externalize the solution steps taken by a problemsolver by following the path of reasoning from the point at which the data considered relevant were selected, through the concepts and processes utilized, to the intermediate and final responses. The categorization of the response is based on (i) the data selection, (ii) the nature of the implicit and explicit concepts used, and, most importantly, (iii) the way in which the two are integrated. The other technique (Siemon, 1988) combines the mapping procedure with an interview in which the student's description of what he/she is doing while working on the problem is linked with the map. In working with this method on an experimental basis, it has been found that when the student's affective reactions at various stages of the process are recorded, they are often as revealing as the explicit statements of method.

Let us consider an example set in this open-ended format; the sample responses in Figure 2.2 were identified by interview.

Find the value of Δ in the following statement:

$$(72 \div 36) = (72 \times 9) \div (\Delta \times 9)$$

Prestructural responses

> "Have not done ones like that before, so I can't do it."
> "Don't want to do it."

Comment: Both respondents indicate that they are unwilling to engage in the task.

Unistructural responses

> "36 – because there is no 36 on the other side."
> "2 – because 72 ÷ 36 = 2."

Comment: Both responses take only one part of the data into account. The first response shows a low level "pattern completion" strategy. The second response shows one closure and then an ignoring of the remainder of the item.

Multistructural response

$$2 \times 9 = 18 \qquad\qquad 648 \div (\Delta \times 9)$$
$$648 \div ? = 2 \text{ that is, } 324$$
$$\text{looking for } 18 \ (2 \times 9)$$

$$\text{Hence } 324$$

Comment: This response incorporates a series of arithmetical enclosures to reduce the complexity and to focus on "Δ." However, the students appear unable to keep the overall relationship in mind throughout the closure sequences and lost in a "maze" of their own creation.

Relational response

$$2 \times 9 = 18 \qquad\qquad 648 \div (\Delta \times 9)$$
$$648 \div 9 = 72$$
$$\text{then } 72 \div 4 = 18$$

$$\text{Hence } 4$$

Comment: This response also involves a sequence of arithmetical closures, but the students are able to keep the relationships within the statement in mind and thus successfully solve the problem.

Extended abstract response

First step involves obtaining an overview of the relationships between the numbers and operations involved, for example:

$$(72 \div 36) \times 9 = (72 \times 9) \div (\Delta \times 9)$$

The pattern suggests something akin to the "distributive" property—this hypothesis is tested out thus:

$$\frac{a}{b} \times y = \frac{a \times y}{b}$$

This immediately solves the problem (without necessity for closure) as follows:

$$(72 \div 36) \times 9 = (72 \times 9) \div 36 = (72 \times 9) \div (4 \times 9)$$

$$\text{Hence } 4$$

Comment: This response shows the following characteristics:

1. Focusing on the relationships between the operations and the numbers rather than regarding the operations as instructions to close;

2. a hypothesis suggested by the data;

3. avoiding closures wherever possible as these change the form of the statement and "hide" the original relationship.

Figure 2.2. Open mathematics item (Biggs & Collis, 1982, pp. 83–84).

SOLO taxonomy: Closed format

This form was developed initially for use in testing mathematical problem solving (Collis, 1982; Collis, Romberg, & Jurdak, 1986) by combining the superitem technique devised by Cureton (1965) with the cycle of learning notion from the SOLO taxonomy. It requires the writing of an item stem that provides data for four questions devised in such a way that each requires an ability to respond at one of the SOLO levels: unistructural, multistructural, relational, or extended abstract. The basic criteria for designing the questions are as follows: (i) unistructural—use of one obvious piece of information coming directly from the stem; (ii) multistructural—use of two or more discrete closures directly related to separate pieces of information in the stem; (iii) relational—use of two or more closures directly related to an integrated understanding of the information in the stem; and (iv) extended abstract—use of an abstract general principle or hypothesis which is derived from (or suggested by) the information in the stem.

A method of construction and certain psychometric analysis of the data gathered (Wilson & Iventosch, 1985) for mathematical problem solving has been reported (Collis, Romberg, & Jurdak, 1986) (see Figure 2.3).

Assessment in Great Britain

Some of the recent innovative approaches to assessment in Great Britain may prove useful. The Assessment of Performance Unit (APU), similar to the National Assessment of Educational Progress in the United States, was commissioned to prepare a national profile on the educational achievement of children. The work of the APU is geared toward causing educational change by having assessment procedures precipitate curricular change. The direction of change is essentially that outlined as desirable by the Cockroft Commission (Committee of Inquiry into the Teaching of Mathematics in Schools, 1982), which advocated, among other things, links with other curricular areas, practical work, the importance of language, a diagnostic approach to testing (cf. Bell, 1985), mathematics for the majority, a graduated assessment, and records of progress. In the process, the APU gave completion tests to a large number of stu-dents. One facet consisted of a matrix-sampling approach organized around a content-by-behavior matrix to which had been added a third dimension that addressed understanding, practical application, problem solving, and attitudes. The third dimension, involving the more innovative efforts, was assessed separately by sending test booklets to small samples.

The APU's assessment methods for the practical and problem-solving parts (Foxman, 1985; Foxman & Mitchell, 1983) are a combination of pencil-and-paper answers to complex and realistic situations and practical assessment with manipulatives in a diagnostic assessment interview (e.g., Denvir & Brown, 1985, 1986; Joffe, 1985). The situational questions are largely analogous to the superitem approach in that there is a problem stem with considerable information, followed by a series of increasingly complex questions. Answers can range from the simple to the complex. Diagnostic interviewing of a small sample of students engaged in a practical test is conducted according to a script, but with some flexibility for clarification, limited prompting, or amended answers. Responses are checked against a precoded list, but unanticipated answers are recorded in detail. The result is valuable insight into students' mathematical thinking (Burstall, 1986).

It has been argued that intelligence must now be regarded as multifaceted (Walters & Gardner, 1985) and susceptible to improvement. Therefore, methodologies and instruments are needed that do more than produce a crude terminal score purporting to summarize years of achievement.

A number of strategies have been tried in Great Britain which essentially link internally created portfolios of student work with teacher assessment and external moderation. SMP, for example, would have students engage in a number of extended phases of work over the course of a year, each lasting approximately two weeks. SMP offers some starting points for project work, but leaves choice open to the students. The logs and reports would be assessed by teachers according to guidelines provided by SMP. Teachers in a school or region would meet at intervals to discuss their grading in a collegial, moderation process.

This is a machine that changes numbers.
It adds the number you put in three times
and then adds 2 more. So, if you put in 4,
it puts out 14.

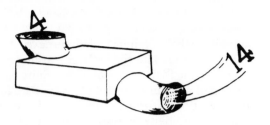

U. If 14 is put out, what number was put in?

Answer _____

Answer: 4

Comment: Students have to understand the problem well enough to be able to close on the correct response which is displayed in the stem.

✎ ✏ ✐

M. If we put in a 5, what number will the machine put out?

Answer _____

Answer: 17

Comment: Students need to comprehend the set problem sufficiently to be able to use the given statements as a recipe and thus perform a sequence of closures which they do not necessarily relate to one another.

✎ ✏ ✐

R. If we got out a 41, what number was put in?

Answer _____

Answer: 13

Comment: An integrated understanding of the statements in the problem is necessary to carry out a successful solution strategy in this case. Correct solutions may involve working backwards or carrying out a series of approximation trials. It should be noted that the solution requires only data-constrained reasoning in that no abstract principles need to be invoked.

✎ ✏ ✐

E. If "X" is the number that comes out of the machine
when the number "Y" is put in, write down a formula which
will give us the value of "Y" whatever the value of "X."

Answer _____

Answer: $Y = \dfrac{X-2}{3}$

Comment: A correct response involves extracting the relationships from the problem and setting them down in an abstract formula. It involves using the information given in a way quite different from that of the lower levels.

Figure 2.3. Closed mathematics item (adapted from Collis, Romberg & Jurdak, 1986).

Three of the students' projects would be submitted for external assessment. At least one of the three would be a practical project, such as designing packaging for candies. A second must be an investigational project in which the student engages in mathematically reflective activity. The internally assessed (school), but externally moderated (examination board) record is intended to promote many of the practices attempted in pilot efforts (Wharmby, 1986): (i) a modular approach, (ii) practical work, (iii) extended project work, (iv) written assignments, (v) oral assessment, (vi) written assessment, (vii) assessment as an integral part of the learning process, (viii) greater involvement of the teacher in the assessment process, (ix) a cumulative profiling of students' mathematical achievement, and (x) an implicit intent to send all students — not just the brightest and most mathematically able — into the adult world with some mathematical understanding and confidence.

Graded Assessment in Mathematics (GAIM) also stresses assessment tasks that are good instructional tasks and therefore have curricular validity. The curriculum is divided into progressive levels (Brown, 1986) determined by the facility hierarchies identified in the Concepts in Secondary Mathematics and Science Project (Hart, 1980). Practical and investigative activities in this project are shorter (20 minutes to 1.5 hours). They start from specific but open-ended activity cards and are linked to student and class profiles.

In summary, the thrust of the effort in Great Britain is toward a much wider variety of teaching and learning strategies, with the assessment process regarded as a catalyst. Proposals for national assessment include a score-weighted battery of written tests, performance tests, tests of mental arithmetic, an oral interview, and a review of project work. Ongoing experiments address integrated assessment tasks that simultaneously assess students in a performance task that includes science, mathematics, and written and verbal communication. These are multifaceted strategies that have the potential for providing a more flexible and much more detailed picture of student achievement that simultaneously contributes to their understanding of the subject.

Summary and Conclusion

Traditional assessment practices have consistently used a content-by-behavior matrix as their theoretical framework and relied heavily on independent, multiple-choice items. Cost efficiency almost eradicated other approaches to group testing. However, the mathematical, psychological, sociological, and pedagogical theories embedded in the model are, quite simply, inadequate. Consequently, it is important to replace the matrix model with one more capable of handling complexity and one that will stimulate change. Unfortunately, the cohesive power of the matrix model exerts a powerful influence which subliminally impedes change.

It is essential that the new model be powerful and have both tight internal coherence and congruence with current trends in mathematics, science, and society. It is also important that the key indicators and instruments for measuring be equally coherent and congruent, the cohesive force being purpose — namely, the development of each student's mathematical power.

Because the intent is to assess the creation of knowledge and the processes involved rather than to measure the extent to which students have acquired a coverage of the field of mathematics, a much wider variety of measures, many of them qualitative, are needed. Considerable effort is needed to find instruments adequate for the purpose.

Only a few of a wide variety of approaches to assessment have been discussed here. They were selected as representative of the range of instruments that might form a coherent repertoire. Some are theory driven by recent research in cognition which would support the new world view, and others are the result of practical insights by educators with a new world view of mathematics. Both approaches are appropriate at this stage. The urgent need is for a much greater variety of learning and assessment tasks, a coherent body of tools that will precipitate curricular change.

References

Bell, A. W. (1985). Some implications of research on the teaching of mathematics. In A. Bell, B.

Low, & J. Kilpatrick (Eds.), *Theory, research, and practice in mathematical education* (pp. 61-79, 5th International Congress on Mathematical Education, Adelaide, South Australia, 1984). Nottingham, England: Shell Centre for Mathematical Education, University of Nottingham.

Berlak, H. (1985, October). Testing in a democracy. *Educational Leadership, 43*(2), 16-17.

Biggs, J. B., & Collis, K. F. (in press). Multimodal learning and the quality of intelligent behavior. In H. Row (Ed.), *Intelligence: Reconceptualization and measurement*. Melbourne, Australia: ACER.

Biggs, J. B., & Collis, K. F. (1989). Towards a model of school-based curriculum development and assessment using the SOLO taxonomy. *Australian Journal of Education, 33*(2).

Biggs, J. B., & Collis, K. F. (1982). *Evaluating the quality of learning: The SOLO taxonomy (structure of the observed learning outcome)*. New York: Academic Press.

Bloom, B. S. (Ed.). (1956). *Taxonomy of educational objectives: The classification of educational goals. Handbook 1: Cognitive domain*. New York: Longman.

Brown, M. (1986). Developing a model to describe the mathematical programs of secondary school students (11 – 16 years): Findings of the graded assessment in mathematics project. In *Psychology of Mathematics Education: Proceedings of the Tenth International Conference* (pp. 135-140). London: The University of London Institute of Education.

Burstall, C. (1986). Alternative forms of assessment: A United Kingdom perspective. *Educational Measurement: Issues and Practice, 5*(1), 17-22.

Cambridge Institute of Education. (1985). *New perspectives on the mathematics curriculum: An independent appraisal of the outcomes of APU mathematics testing, 1978 – 1982*. London: Her Majesty's Stationery Office.

Carpenter, T. P., Coburn, T. G., Reys, R. E., & Wilson, J. W. (1978). *Results from the first mathematics assessment of the national assessment of educational progress*. Reston, VA: The National Council of Teachers of Mathematics.

Carpenter, T. P., Corbitt, M. K., Kepner, H. S., Jr., Lindquist, M. M., & Reys, R. B. (1981). *Results from the second mathematics assessment of the national assessment of educational progress*. Reston, VA: the National Council of Teachers of Mathematics.

Case, R. (1985). *Cognitive development*. New York: Academic Press.

Chick, H. L. (1988). Student responses to a polynomial problem in light of the SOLO taxonomy. *Australian Senior Mathematics Journal, 2*, 91-110.

Chick, H. L., Watson, J. M., & Collis, K. F. (1988). Using the SOLO taxonomy for error analysis in mathematics. *Research in Mathematics Education in Australia*, 34-37.

Collis, K. F. (1982). The SOLO taxonomy as a basis of assessing levels of reasoning in mathematical problem solving. In *Proceedings of the Sixth International Conference for the Psychology of Mathematical Education*. Antwerp, Belgium: University of Antwerp.

Collis, K. F., & Romberg, T. A. (1981). Cognitive functioning, classroom learning, and evaluation: Two projects. In *Proceedings of the Fifth International Conference for the Psychology of Mathematical Education* (pp. 64-77). Grenoble, France: University of Grenoble.

Collis, K. F., Romberg, T. A., & Jurdak, M. E. (1986). A technique for assessing mathematical problem-solving ability. *Journal for Research in Mathematics Education, 17*(3), 206-221.

Collis, K. F., & Watson, J. M. (1989). A SOLO mapping procedure. In *Proceedings of the Thirteenth International Conference for the Psychology of Mathematical Education*. Paris: University of Paris.

Cureton, E. E. (1965). Reliability and validity: Basic assumptions and experimental designs. *Educational and Psychological Measurement, 25*, 326-346.

Committee of Inquiry into the Teaching of Mathematics in Schools. (1982). *Mathematics counts* (The Cockroft Report). London: Her Majesty's Stationery Office.

Demetriou, A., & Efklides, A. (1985). Structure and sequence of formal and postformal thought: General patterns and individual differences. *Child Development, 53*, 1062-1091.

Denvir, B., & Brown, M. (1985). Understanding of number concepts in low attaining 7-9 year olds: Part I. Development of descriptive framework and diagnostic instrument. *Educational Studies in Mathematics, 17*(1), 1-22.

Denvir, B., & Brown, M. (1986). Understanding of number concepts in low attaining 7-9 year olds: Part II. The teaching studies. *Educational Studies in Mathematics, 17*(2), 51-72.

Eisenberg, T. (1975). Behaviorism: The bane of school mathematics. *Journal of Mathematical Education, Science, and Technology, 6*(2), 163-171.

Fischer, K. (1980). A theory of cognitive development: The control and construction of hierarchies of skills. *Psychological Review, 57*, 477-531.

Foxman, D. (1985). Mathematics assessment framework. In *Mathematics report for teachers: 1 (DRAFT)*. London: Assessment of Performance Unit, Department of Education and Science.

Foxman, D. D., Cresswell, M. J., & Badger, M. E. (1981). *Mathematical development. Primary survey report no. 2: Report on the 1979 primary survey from the National Foundation for Educational Research in England and Wales to the Department of Education and Science, the Department of Education for Northern Ireland and the Welsh Office.* London: Her Majesty's Stationery Office.

Foxman, D., & Mitchell, P. (1983, November). Assessing mathematics: 1. APU framework and modes of assessment. *Mathematics in Schools, 12*(4), 1-5.

Frederiksen, N. (1984). The real test bias. *American Psychologist, 39*(3), 193-202.

Graded Assessment in Mathematics. (1986, Spring). Newsletter 4. London: King's College/University of London/ILEA.

Halford, G. S. (1982). *The development of thought.* Hillsdale, NJ: Lawrence Erlbaum.

Hart, K. (1980). *Secondary school children's understanding of mathematics. A report of the mathematics component of the concepts in secondary mathematics and science programme.* London: Mathematics Education/Centre for Science Education, Chelsea College, University of London.

Hilton, P. (1981). Avoiding math avoidance. In L. A. Steen (Ed.), *Mathematics tomorrow* (pp. 73-82). New York: Springer-Verlag.

Joffe, L. (1985). *Practical testing in mathematics at age 15.* London: Assessment of Performance Unit, Department of Education and Science.

McLean, L. D. (1982). Achievement testing — Yes! Achievement tests — No. *E + M Newsletter, 39,* 1-2.

Mounod, P. (1985). Similarities between developmental sequences at different age periods. In I. Levin (Ed.), *Stage and structure.* New York: Ablex.

National Assessment of Educational Progress. (1981). *Mathematics objectives: 1981-82 assessment* (No. 13-MA-10). Princeton, NJ: Educational Testing Service.

National Commission on Excellence in Education. (1983). *A nation at risk: The imperative for educational reform.* Washington, DC: US Government Printing Office.

National Council of Teachers of Mathematics. (1989). *Curriculum and evaluation standards for school mathematics.* Reston, VA: Author.

Reich, R. B. (1983). *The next American frontier.* Harmondsworth, England: Penguin Books.

Resnick, L. B. (1987). *Education and learning to think.* Washington, DC: National Academy Press.

Romberg, T. A. (1985, December). *The content validity for school mathematics in the US of the mathematics subscores and items for the second international mathematics study.* Paper presented for the Committee on National Statistics, National Research Council of the National Academy of Sciences. Madison, WI: Wisconsin Center for Education Research.

Romberg, T. A., & Kilpatrick, J. (1969). Appendix D. Preliminary study on evaluation in mathematics education. In T. A. Romberg & J. W. Wilson (Eds.), *The development of tests. NLSMA report no. 7* (pp. 281-298). Stanford, CA: School Mathematics Study Group.

Romberg, T. A., & Wilson, J. W. (1969). *The development of tests. NLSMA report no. 7.* Stanford, CA: School Mathematics Study Group.

Romberg, T. A., Zarinnia, E. A., & Williams, S. R. (1989). *The influence of mandated testing on mathematics instruction: Grade 8 teachers' perceptions.* Madison, WI: National Center for Research in Mathematical Sciences Education.

Scheffler, I. (1975, October). Basic mathematical skills: Some philosophical and practical remarks. In *The NIE Conference on basic mathematical skills and learning. Volume I: Contributed position papers* (pp. 182-189), Euclid, Ohio. Los Alamitos, CA: SWRL Educational Research and Development.

Siemon, D. E. (1988). The role of metacognition in children's mathematical problem solving: A question of construct and methodology. In *Proceedings of the 11th annual conference of Mathematics Education Research Groups of Australia.* Melbourne, Australia: Deakin University.

Thorndike, E. L. (1904). *An introduction to the theory of mental and social measurements.* New York: Teachers College, Columbia University.

Tyler, R. W. (1931). A generalized technique for constructing achievement tests. *Educational Research Bulletin, 8,* 199-208.

Walters, J. M., & Gardner, H. (1985). The development and education of intelligences. In F. Link (Ed.), *Essays on the intellect* (pp. 1-21). Alexandria, VA: Association for Supervision and Curriculum Development.

Weinzweig, A. I., & Wilson, J. W. (1977, January). *Second IEA mathematics study. Suggested tables of specifications for the IEA mathematics tests* (Working paper 1). Wellington, New Zealand: IEA.

Westbury, I. (1980, January). Change and stability in the curriculum: An overview of the questions. In *Comparative studies of mathematics curricula: Change and stability, 1960–1980* (pp. 12-36). Proceedings of a conference jointly organized by the Institute for the Didactics of Mathematics (IDM) and the International Mathematics Committee of the Second International Mathematics Study of the International Association for the Evaluation of Educational Achievement (IEA). Bielefeld, West Germany: Institut fur Didaktik der Mathematik der Universitat Bielefeld.

Wharmby, D. (1986). *SEC-sponsored development project on practical mathematics: A review of research projects in mathematics.* Manchester, England: Joint Matriculation Board.

Wilson, M., & Iventosch, L. (1985, April). *Using the partial credit models to investigate responses to structured subtests.* Paper presented at the annual meeting of the American Educational Research Association, San Francisco.

Zarinnia, E. A., & Romberg, T. A. (1987). A new world view and its impact on school mathematics. In T. A. Romberg & D. M. Stewart (Eds.), *The monitoring of school mathematics: Background papers. Vol. 1: The monitoring project and mathematics curriculum* (pp. 21-61). Madison, WI: Wisconsin Center for Education Research.

3

Power Items and the Alignment of Curriculum and Assessment

TEJ PANDEY

The declining test scores on NAEP, SAT, Iowa, and several state testing programs from the mid-1960s to mid-1970s have resulted in many studies and commission reports, such as *A Nation At Risk, The Paideia Proposal, A Place Called School, A Study of High Schools, Educating Americans for the 21st Century,* and *The American High School.* These reports generated concern for better education, resulting in the "excellence movement." Many states and school districts across the nation have set high expectations and goals for all learners and have devised programs for helping students and schools in ways that allow them to reach those high expectations. The ideas from effective school research have become the hallmark to promote excellence. Among many effective school practices, the one that is important to the discussion in this chapter is the organizing of clear and visible goals around which instruction can be continually targeted and implementing assessment and indicator programs to monitor the progress of students as well as of school systems. What is curriculum alignment and what are some of the issues related to curriculum, instruction, and assessment (CIA) alignment in the setting of the current reform movement?

This chapter will discuss issues and strategies in planning assessment instruments in mathematics that would support the reform. In particular, the discussion focuses on the nature of the test instruments most suitable for large-scale accountability assessment programs. It is recommended that "power items" be used for such purposes. Suggestions are offered for defining content domains and item specifications and for constructing power questions in the context of CIA alignment.

Curriculum Alignment

The NCTM *Standards for School Mathematics* (1989) states that "In assessing students' learning, assessment methods and tasks should be aligned with the content and instructional goals of the curriculum." In a broader sense, the curriculum is composed of goals and objectives, or intended curriculum, instruction, and assessment. When all three match—that is, instruction and assessment focus on stated objectives—then the effects of schooling are usually both understandable and impressive. In other words, instruction and assessment should both be derived from the objectives, and instruction and assessment must be designed to support each other. The greater the mismatch between the tests and instruction with the intended curriculum, the more uncertain we are about the instructional needs of the students and the effectiveness of the instructional program. To derive meaningful inferences, instruction, as well as assessment content and procedures, should be derived from the goals of the intended curriculum.

A lack of alignment occurs because either the instruction or the tests or both are not derived from the goals and objectives. "Tests drive the curriculum" and "textbooks are the curriculum" are two well-known truisms. As propositions, they are oversimplified, yet they reflect the dismay that follows when tests and/or texts are not aligned with a school's intended curriculum.

At least in the area of mathematics, there is evidence showing considerable mismatch between the goals of instruction and the content of textbooks, as well as in the breadth and procedures of the test instruments most commonly used. For example, in 1985, the California State Department of Education called for textbook adoption proposals to be submitted using the criteria that would assure that the content and approach described in the *Mathematics Framework for California Public Schools: Kindergarten Through Grade Twelve* (1985) and *Mathematics Model Curriculum Guide: Kindergarten Through Grade Eight* (1987) would be reflected in the new submissions. The California Curriculum Commission, which screens the textbooks for adoption, found that the textbooks did not meet the minimum selection criteria and allowed an extra year for publishers to revise the textbooks. New textbooks have now been adopted; however, the revisions typically resulted in replacing the instructional approach in only 10 percent of the lessons, leaving a major portion of the series still inconsistent with the intent of the *Framework* (Denham, 1988).

There is also evidence that many of the standardized tests in use fail to match the breadth and approach of the intended curriculum. In a review of content alignment of the *California Mathematics Framework* with that of a sample (nonscientific) of standardized tests from seven publishers, the California Mathematics Council (1986) found that all tests used only multiple-choice type of questions, which predominantly assessed students proficiencies in the knowledge of facts and execution of algorithms; only superficial attention was given to items requiring students to solve problems, reason, or communicate. The California Mathematics Council profiled a typical sixth grade test of 100 questions as follows: number, 67 questions; measurement, 6 questions; geometry, 7 questions; patterns and functions, no questions; probability and statistics, 8 questions; logic, no questions; algebra, 5 questions; word problems, 8 questions; and problem-solving strategies, 1 question.

Changing Role of Standardized Tests

It is interesting to follow the role of testing along with societal and educational concerns. Developed primarily to identify students in need of special help, tests today are seen as an instrument for educational improvement. In recent history, the usefulness of objectively scorable standardized tests for mass testing was first demonstrated during World War I with the use of the Army Alpha for classifying military personnel. The primary purpose of these tests was to select a smaller number of applicants from all those who applied. In 1926, the College Board administered the first Scholastic Aptitude Test (SAT) in the form of essays to approximately 8,000 candidates for admission to colleges. Eleven years later, in 1937, multiple-choice tests replaced the College Board's essay tests (Angoff & Dyer, 1971).

Objective multiple-choice achievement tests began to show a rapid growth in the mid-1950s, with the pioneering of highly sophisticated test-scoring machinery by Lindquist (1954). Today, no one really knows how many standardized tests are administered each year; a rough estimate is that about 30 to 35 million tests are administered annually in the United States. Much of this testing results from the administration of the standardized achievement tests such as Iowa, SAT, CTBS, CAT, and the testing programs, mandated or otherwise, in all 50 states.

Popham (1983) points out that testing had little impact on instruction in the 1950s and early 1960s. There was no media coverage of changes in mean test scores, and scores were used mainly for ability grouping and to identify students who needed help. The situation began to change in 1965 when the Elementary and Secondary Education Act (ESEA) was passed. The act required that certain teaching programs funded by ESEA be evaluated, and it turned out that future funding depended at least in part on the outcomes of the evaluations. As Popham states, "Tests were being employed to make keep-or-kill decisions about educational programs. Big dollars...were riding on the results of achievement tests...the days of penny-ante assessment were over" (p. 23).

Pressure to improve test performance increased during the 1970s when test data showed that attainment of knowledge and skills was declining (Womer, 1981); the National Assessment of Educational Progress (NAEP, 1982) reported decrements in performance; declines in College Board scores attracted nationwide attention; and the means of test scores from the then-existing state programs decreased progressively. Some state legislatures responded by passing laws requiring the use of competency tests for making important educational decisions, such as the granting of a high school diploma. The effect of all this was to increase efforts of schools to prepare students to take the tests. As Popham put it, teachers and school administrators realized that "although students might be the immediate targets...next in line could be educators themselves" (p. 24).

The importance of test scores seems to be rising continuously for schools, districts, states, and the country as a whole. Beginning in 1985, the Secretary of Education has displayed annual educational accomplishments on a state-by-state basis on the wall chart. Besides such factors as the expenditure per pupil and the class size, the wall chart displays the composite SAT scores for each state. A recent report by Alexander and James (1987) recommends that NAEP be expanded, and beginning in 1992, NAEP will collect and report achievement test results on a state-by-state basis.

At this stage in the current reform movement, states such as California, New Jersey, Connecticut, Ohio, and Colorado are engaged in well-structured, effective school reform strategies. In California, for example, Honig (1985) calls for a strategy of creating a vision that informs local effort, legitimizes standards of excellence, maximizes autonomy for units operating under that vision and those standards, recognizes competence, and holds everyone accountable for measurable results. Honig translated his strategy into such actions

as establishing an agreed-upon definition of curriculum reflecting the broad goals, initiating a textbook selection process that reflects instructional standards, improving selection and training of principals, upgrading teacher preparation, seeking community support, defining the concept of effectiveness, setting targets for progress, and devising a broadly conceptualized accountability program. Each year, the *California Performance Reports* (1985, 1986, 1987) publish the progress on a school-by-school basis, using several factors related to student learning as well as the achievement scores based upon tests developed by the California Assessment Program.

Influence of Testing on Instruction

Tests have traditionally been conceptualized as measures of student achievement, much as a weather barometer measures atmospheric pressure. However, unlike barometers, test instruments can affect instruction for better or for worse, depending on the ways that they are constructed and used. Lerner (1987) has reviewed several major studies dealing with the influence of time and effort on school achievement. She summarizes her review by saying:

Overall, data from many converging sources suggest that time on task is a major determinant of educational outcomes and that external demand levels have a significant effect on the amounts of time and effort students put into their schoolwork. Inevitably, minimum competency testing programs tend to increase the amounts of time and effort spent working on basic skills for at least two reasons. First, the tests themselves constitute a set of external demands.... Second...teachers usually "teach to the tests" to at least a limited degree. (pp. 1061–1062)

In a recent study, Cannell (1988) has noted that almost all of the 50 states have shown consistent improvement on the publisher's standardized tests and that the average student remains above the publisher's norms set for the standardizing year. Although improvement in student achievement is much to be desired, and the use of tests to achieve that outcome has even been promoted (Popham, 1987), it is tempting to hypothesize that the reliance on commonly available objective tests, particularly the multiple-choice tests, to provide evidence of improvement may lead to a distortion in education. The distortion may occur because such tests may lead to an increased emphasis on material that is easily tested but is less important educationally and may lead to a decrease in efforts to teach other important abilities that are difficult to measure. Reports from NAEP (1982) and other testing programs show that performance on items measuring "basic skills" is improving; however, there is a decrease in performance on items that measure more complex cognitive skills. Although there could be many reasons to account for the performance on NAEP tests, Frederiksen (1984) has pointed out that "the possibility must be considered that the mandated use of minimum competency tests, which use the multiple-choice format almost exclusively, may have discouraged the teaching of abilities that cannot easily be measured with multiple-choice items."

The use of tests and test results to measure outcomes of the reform effort is criticized by Wise (1988). He states that:

By mandating educational outcomes through standardized tests, content through curriculum alignment, and teaching methods through teacher evaluation criteria, states set in motion a chain of events that alter educational ends and means.... A standardized test would set the educational objectives for the teacher. Curriculum alignment would insure that the teacher would cover the material to be tested.... Less obvious, however, are the distortions introduced into the curriculum by testing. Some teachers begin to emphasize the content that they know will appear on the test. They begin to teach in a format that will prepare students to deal with the content as it will be tested. Some even teach items that are likely to appear on the test. Meanwhile, the rest of the curriculum is deemphasized. (p. 330)

Using Tests to Improve Education

Although we do not believe that tests should be used to push for student learning, in situations where tests drive the instruction, the ill effects of testing can be minimized or even be used to promote good instructional practices through proper content coverage and test procedure. Frederiksen (1984) has suggested that in "symbolic" test design, test instruments must reflect the entire domain of educational goals and develop questions that are most meaningful to students and helpful in improving the educational process. Honig (1985) proposed the use of "power items" for accountability-type tests. The power items assess important educational outcomes developed over a period of instructional sequence, emphasize understanding, integrate a number of ideas, and serve as examplars of good instructional practices.

The following sections describe the strategies useful for developing the power items. They not only help align assessment with the goals of the curriculum; they also could be useful for surprisingly good instructional practices. The discussion will emphasize that (i) domain specification should be derived solely from the consensus of discipline leadership rather than analysis of textbook or common curriculum in schools, (ii) the content-by-process matrix must incorporate the knowledge acquisition processes and be viewed as having no boundaries at the time of item writing, to encourage interconnectedness of mathematical ideas, (iii) item writing should be done as a team process to encourage writing questions in hard-to-assess areas, such as understanding communicating, generalizing, formulating, and problem solving, and (iv) besides multiple-choice, other modes of assessment—such as open-ended, situational investigations, and portfolios—be used.

Domain specifications: Consensus vs. leadership

Mehrens (1984) summarizes how the domains of the publishers' standardized tests are generally determined. The first step in delineating the test content is to identify what is being taught nationwide by analyzing published textbook series, syllabuses, outlines of objectives, and other curricular materials. Curriculum experts are also consulted regarding what trends the curriculum might follow in the future. From a compilation of this information, publishers then "distill the common essence of that set.... There is a reasonably commonly taught subdomain and indeed each test and textbook publisher attempt to sample from that subdomain." Mehrens adds that:

Because publishers wish to tap a common curricular subdomain, content not tested will probably occur more than content not taught. This is certainly going to be true if one does a study of mismatch at the item level, because a 50-item test can only test at most 50 specific objectives!

However, we need a much broader conception of what a test is if we are to use test information to improve educational outcomes. This does not mean that the previously described tests constructed using objective description have no usefulness; nonetheless, if such tests are used as examplars and guides by a vast majority of teachers to steer the course of their instruction, curriculum distortion will result. For tests used in accountability programs, the test objectives must be derived from broad curriculum goals and framework. The objectives should be derived from a consensus of professional leadership in subject matter as opposed to a common denominator of content in textbooks or content actually taught. In most cases, the tests will probably be broader and deeper than the instructional practices in schools.

The two publications likely to be most prominent as a basis of defining objectives for mathematics learning in schools in the next 10 to 15 years are *Curriculum and Evaluation Standards for School Mathematics* (1989), published by the National Council of Teachers of Mathematics, and *Mathematics Curriculum for the 21st Century* (1985), published by the Mathematical Sciences Education Board. As part of its reform, the California State Department of Education published *Mathematics Framework for California Public Schools: Kindergarten Through Grade Twelve* (1985), *Model Curriculum Standards: Grades Nine Through Twelve* (1985), and *Mathematics Curriculum Guides: Kindergarten Through Grade Eight* (1987). These documents emphasize the role of communicating, examining, representing, transforming, hypothesizing, proving, applying, and solving problems in learning

mathematical skills. The new tests developed by the California Assessment Program are based upon the goals and recommendations of these documents rather than on a consensus approach leading toward the least common denominator of content coverage.

Item Specifications

Tests used for accountability programs not only require the breadth and depth of the content but also require an approach other than Bloom's (1956). Typically, item specification starts with a content-by-process matrix, then, for each cell of the matrix, domains and subdomains are specified. As recommended by Popham (1980) and Millman (1974), the description of specifications ensures that an item is a true reflection of the intended skill to be measured.

The approach of delineating the content-by-process matrix, and then describing the domains within the cells, results in very narrowly defined, well-structured—in fact, too structured—test questions. The narrow structures result in test questions that assess a minuscule piece of information, roughly one-minute-per-question tasks. If teachers were to use such test questions or the test specifications as a model to improve test scores, they would be likely to reinforce a fragmented, nonintegrated, multiple-choice type of instruction.

Tests built on such specifications are likely to improve the educational process only in the narrow sense that they perpetuate the teaching of what is measured and make it more effective. The teachers are unlikely to emphasize connections between the various concepts, which is necessary to develop student cognition at a higher level.

In order to influence instruction in a positive way, we believe test questions should be ill-structured rather than well-structured. Simon (1978) defines an ill-structured problem as one that is complex, without definite criteria for determining when the problem is solved, without all the information needed to solve the problem, and without a "legal move generator" for finding all the possibilities at each step in solving the problem. Frederiksen

(1984) also justifies the role of ill-structured problems in accountability-type tests and states that:

Most of the important problems one faces in real life are ill-structured, as are the really important social, political, and scientific problems in the world today. But ill-structured problems are not found in standardized achievement tests. It would be unfair, if schools do not teach students how to solve ill-structured problems. To make it fair, schools would have to teach the appropriate problem-solving skills.... Ability to solve ill-structured problems is just one example of a desirable outcome of education that is not tested and not taught. (p. 199)

Frederiksen proposes direct observation of behavior on situational tasks that are important in school learning. Recognizing that the development of such tests is not simple, Frederiksen (1984) states that:

There are problems of discovering what are the salient aspects of performance in carrying out a particular task and in identifying the cognitive processes that it requires. There are problems of scoring... and problems concerned with time and cost of testing that may require compromises (p. 199).

We propose "softer" domain specifications, serving more as a guide to assure coverage, rather than "rigid" specifications casting look-alike items. Also, rather than using the conventional content-by-process matrix-based taxonomy, we recommend a taxonomy that is derived from the theories of knowledge acquisition, such as the theory of information processing (e.g., see Anderson, 1981; Glaser, 1978; Resnick, 1987). Some of the cognitive processes that have been identified have to do with development of communicating; organizing information; representing problems internally; recognizing patterns; understanding, reasoning, and thinking; using strategies and heuristics in problem solving; and metacognition. These knowledge acquisition processes should be allowed to combine freely and effectively with domains of content categories, such as number, algebra, geometry, measurement, and probability and

statistics in many possible ways.

Rather than relying upon the procedure of test specification and item writers writing the items, this proposal means working directly, as a team, with mentor teachers, curriculum experts, psychometricians, and cognitive psychologists to develop test questions. Of course, at the heart of this team are the adventuresome mentor teachers who have discovered and practiced successful simulated situations formally or informally in their classrooms. This team would develop a sufficient number and variety of problems simulating situations that require skills worthy of classroom instruction. The team would know that teaching or coaching for such tests would be desirable and would enhance generalization of skills from school to real-life situations. The test questions would go through several scrutinies to assure that all relevant and important objectives of the *Framework* are covered comprehensively and in depth.

In short, this approach of item specification means that the lines dividing a content-by-process matrix are fussy; items are allowed to cut freely across the lines forming the cell of the matrix. Moreover, items are represented more like blobs rather than as dots in the criss-crossed content-by-process matrix:

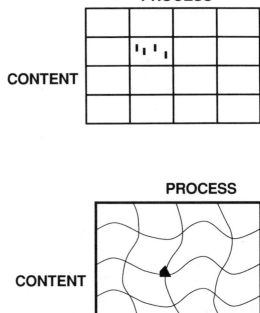

Modes of Assessment

Traditionally, we have relied on objective, multiple-choice questions for classroom or large-scale assessments. Although multiple-choice questions are objective, efficient, and reliable, they are being criticized as lacking the qualities that are significant in improving instruction. For example, commonly used multiple-choice tests convey the idea that curriculum is made up of bits and pieces and that there is only one correct answer to all the problems. There is also a general lack of multiple-choice standardized tests that assess students ability to formulate, reason, conjecture, hypothesize, and communicate.

The *Curriculum Standards for School Mathematics* (1987) recommends that student assessment include a variety of performance measures to ensure that the assessment covers the breadth and depth of the program content. Four general assessment modes are (i) paper-and-pencil objective assessment, (ii) short-answer/essay open-ended assessment, (iii) situational investigation/performance assessment, and (iv) portfolio assessment.

The following section presents the characteristics of multiple-choice and short-answer power questions developed by the California Assessment Program, followed by a few illustrative examples. Other modes of assessment, such as situational problem solving, investigations and performance assessment, are described in other chapters of this volume. [Readers may want to refer to the *Assessment Alternatives in Mathematics* (Stenmark, 1989) for an example of portfolio assessment.]

Power Items

Multiple-choice or short-answer power questions can be characterized as follows: (i) They assess essential mathematical understandings and interconnectedness of mathematical ideas, rather than isolated facts and knowledge; (ii) They represent integration of two or more strands of mathematics, such as number and geometry; (iii) They serve as examplars of good teaching practices that are not likely to distort the teaching and learning process; (iv) They are not directly teachable; however, teaching for them will result in good instruc-

tion; (v) A multiple-choice power item may require three to five minutes to answer compared to a typical multiple-choice item that takes one minute per question. A short-answer, open-ended question may require eight to ten minutes to answer; and (vi) They have face validity such that "good" teachers looking at the questions would feel comfortable and agree that such questions are worthwhile teaching skills.

Examples on page 47 illustrate multiple-choice and open-ended questions. Each question is accompanied by a short description of what the item is designed to measure. The California Assessment Program's *Rationale and Content for the Survey of Academic Skills: Grade Twelve* (1987 draft) describes the specifications with illustrative questions for the test used at the twelfth grade.

References

Alexander, L., & James, H. T. (1987). *The nation's report card: Improving the assessment of student achievement.* Washington, DC: National Academy of Education.

Anderson, J. R. (Ed.). (1981). *Cognitive skills and their acquisition.* Hillsdale, NJ: Lawrence Erlbaum.

Angoff, W. H., & Dyer, H. S. (1971). The Admissions Testing Program. In W. H. Angoff (Ed.), *The College Board Admissions Testing Program* (pp. 1-13). New York: College Entrance Examination Board.

Bloom, B., ed. (1956). *Taxonomy of educational objectives. The classification of educational goals, handbook I: Cognitive domain.* New York: David McKay Co., Inc.

California Assessment Program (1987). *Rationale and Content for the Survey of Academic Skills: Grade 12.* Sacramento: California State Department of Education.

California Mathematics Council (1986). *Standardized tests and the California mathematics curriculum: Where do we stand? A review of content alignment.*

California State Department of Education (1985). *Mathematics framework for California public schools: Kindergarten through grade twelve.* Sacramento.

California State Department of Education (1985). Model curriculum standards: Grades nine through twelve. Sacramento.

California State Department of Education (1987). *Mathematics model curriculum guide: Kinder-*

garten through grade eight. Sacramento.

California State Department of Education (1985, 1986, 1987). *Performance report for California public schools.* Sacramento.

Cannell, J. (1988). The Lake Wobegon Effect Revisited. *Educational measurement: Issues and practice,* Vol. 7, n. 4, pp. 12-15.

Denham, W. (1988). Personal communication.

Frederiksen, N. (1984). The real test bias: Influences of testing on teaching and learning. *American psychologist,* Vol. 39, n. 3, pp. 193-202.

Glaser, R. (Ed.) (1978). *Advances in instructional psychology: Vol. 1.* Hillsdale, NJ: Erlbaum.

Honig, B. (1985). The educational excellence movement: Now comes the hard part. *Phi Delta Kappan,* Vol. 66, n. 10, pp. 675-681.

Lerner, B. (1987). A national census of educational quality: What is needed? *NASSP Bulletin,* Vol. 71, n. 497, pp. 42-44.

Lindquist, E. F. (1954). The Iowa electronic test processing equipment. In *1953 invitational conference on testing problems* (pp. 160-168). Princeton, NJ: Educational Testing Service.

Mathematical Sciences Education Board (1985). *Mathematics Curriculum for the 21st Century.* Washington, DC.

Mehrens, W. A. (1984). National tests and local curriculum: Match or mismatch? *Educational Measurement: Issues and Practices,* Vol. 3, n. 3, pp. 9-15.

Millman, J. (1974). Criterion-referenced measurement. In W. James Popham, *Evaluation in education.* Berkely: McCutchan Publishing Corp.

National Assessment of Educational Progress (1982). Graduates may lack tomorrow's "basics." NAEP *Newsletter, 15,* 8.

National Council of Teachers of Mathematics (1989). *Curriculum and evaluation standards for school mathematics.* Reston, VA.

NCESS Newsletter (1987). National Center for Educational Statistics. Department of Education, Washington, DC.

Popham, J. (1987). The merits of measurement-driven instruction. *Phi Delta Kappan,* Vol. 68, n. 9, pp. 679-82.

Popham, J. (1980). Specifying the domain of content on behaviors. *A guide to criterion-referenced test construction.* Baltimore: John Hopkins University Press.

Popham, W. H. (1983). Measurement as an instructional catalyst. In R. B. Ekstrom (Ed.), *New directions for testing and measurement: Measurement, technology, and individuality in education,* No. 17 (pp. 19-30). San Francisco: Jossey-Bass.

Resnick, L. B. (1987). *Education and learning to think.* Commission on Behavioral and Social Sciences and Education, National Research Council. Washington, DC: National Academy Press.

Simon, H. A. (1978). Information-processing theory of human problem solving. In W. K. Estes (Ed.), *Handbook of learning and cognitive processes: Vol. 5. Human information processing* (pp. 271-295). Hillsdale, NJ: Erlbaum.

Stenmark, J. K. (1989). *Assessment alternatives in mathematics.* Berkeley: University of California.

Wise, A. E. (1988). Legislated learning revisited. *Phi Delta Kappan,* Vol. 69, n. 5, pp. 328-333.

Womer, F. B. (1981). State-level testing: Where we have been may not tell us where we are going. In D. Carlson (Ed.), *New directions for testing and measurement: Testing in the states: Beyond accountability,* No. 10 (pp. 1-12). San Francisco: Jossey-Bass.

Appendix to Chapter 3

Example 1 (multiple choice): Typically students are asked to solve for x when given an equation. In the following question, students are required to see an algebraic representation as mathematization of a real world problem.

Which of the following problems can be solved by using the equation x + 2 = 25?

- A math class started with 25 students. The next day 2 more students enrolled in the class. How many students does this class have now?

- Erin added 2 more books to her collection. If she now has 25 books, how many books did Erin have originally?

- Tim had $25 in his account. A week later he deposited $2 more. How much money does he have in his account now?

- Ann biked 25 km at 2 km per hour. How long did Ann bike?

Example 2 (multiple choice): This question can be done in several ways depending upon the mathematical sophistication of the student. It can be approached purely by trial and error or by trial and error in a systematic way using knowledge of place value.

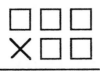

The five digits 1, 2, 3, 4, and 5 are placed in the boxes above to form a multiplication problem. If the digits are placed to give the maximum product, that product will fall between:

- 10,000 and 22,000
- 22,001 and 22,300
- 22,301 and 22,400
- 22,401 and 22,500

Example 3 (multiple choice): The following question combines representation, spatiality, and pattern recognition (multiple). The problem can be done algorithmically or by logical thinking.

A cardboard piece shaped as an equilateral triangle with side 6 cm is rolled to the right a number of times. If the triangle stops so that the letter "T" is again in the upright position, which one of the following distances could it have rolled?

- 24 cm
- 30 cm
- 60 cm
- 90 cm

Example 4 (open-ended) (see pages 48–49): It is generally found that students find it exceedingly difficult to explain, illustrate, describe, or clarify ideas in written or spoken statements. This question is designed to invite dialogue and to find out if students can communicate about geometrical shapes. The question is designed to invite a hierarchy of responses, depending upon the mathematical sophistication of students. They can approach the problem algorithmically or describe in terms of geometric shapes and orientation in space, or in terms of their visualization of other geometric shapes as part of the given shapes.

Example 5 (open-ended) (see pages 50–51): This question is designed to measure students reasoning skills in the context of percents using a real world situation. It also assesses whether students can represent and justify their answers.

California Assessment Program   *Survey of Academic Skills*

OPEN-ENDED MATHEMATICS QUESTION(S)

NAME **M.A.**

Instructions: Use this sheet to answer the questions. Show as much of your work as possible. (In some cases, there may be more than one solution.) Use the reverse side of this sheet if needed.

Imagine you are talking to a student in your class on the telephone and want the student to draw some figures. The other student cannot see the figures. Write a set of directions so that the other student can draw the figures exactly as shown below.

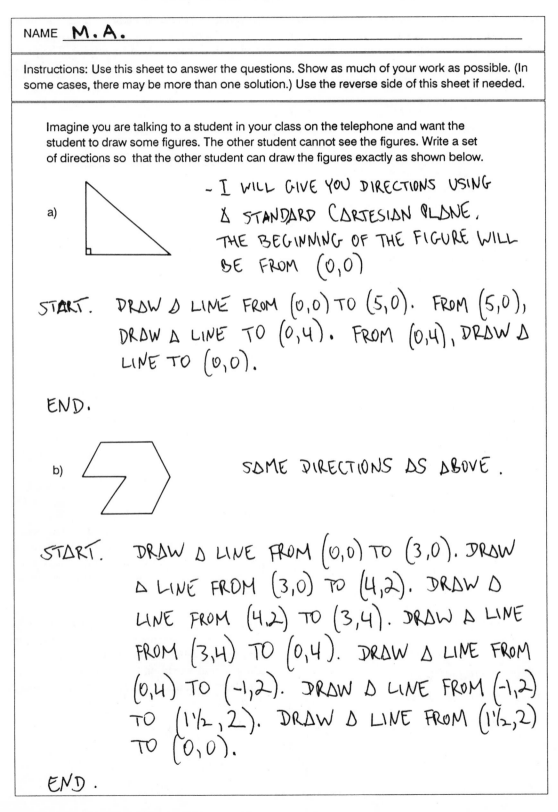

a)

- I WILL GIVE YOU DIRECTIONS USING A STANDARD CARTESIAN PLANE. THE BEGINNING OF THE FIGURE WILL BE FROM (0,0)

START. DRAW A LINE FROM (0,0) TO (5,0). FROM (5,0), DRAW A LINE TO (0,4). FROM (0,4), DRAW A LINE TO (0,0).

END.

b)   SOME DIRECTIONS AS ABOVE.

START. DRAW A LINE FROM (0,0) TO (3,0). DRAW A LINE FROM (3,0) TO (4,2). DRAW A LINE FROM (4,2) TO (3,4). DRAW A LINE FROM (3,4) TO (0,4). DRAW A LINE FROM (0,4) TO (-1,2). DRAW A LINE FROM (-1,2) TO (1½,2). DRAW A LINE FROM (1½,2) TO (0,0).

END.

California Assessment Program

Survey of Academic Skills

OPEN-ENDED MATHEMATICS QUESTION(S)

NAME *P. K.*

Instructions: Use this sheet to answer the questions. Show as much of your work as possible. (In some cases, there may be more than one solution.) Use the reverse side of this sheet if needed.

Imagine you are talking to a student in your class on the telephone and want the student to draw some figures. The other student cannot see the figures. Write a set of directions so that the other student can draw the figures exactly as shown below.

a)

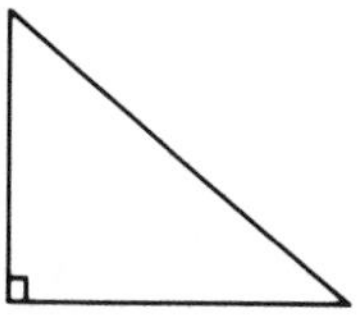

1) draw a 90° angle
2) extend 3 inches up, and 3 inches across
3) join the up part to the across part

b)

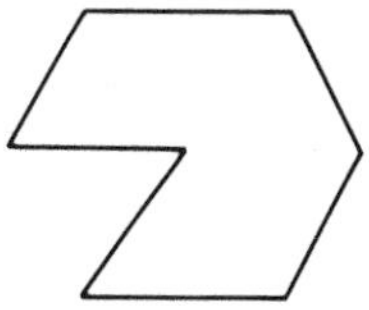

1) Draw a parrallelogram 2 inches by 2 inches and don't fill in the top side
2) draw another and put it on top so it looks like a #7
3) go to the bottom parralelogram, make a line that goes straight across to the left and connect it 2 inches in from the other paralelogram

California Assessment Program   *Survey of Academic Skills*

OPEN-ENDED MATHEMATICS QUESTION(S)

NAME **S.S.**

Instructions: Use this sheet to answer the questions. Show as much of your work as possible. (In some cases, there may be more than one solution.) Use the reverse side of this sheet if needed.

James knows that half of the students from his school are accepted at the public university nearby. Also, half are accepted at the local private college. James thinks that this adds up to 100%, so he will surely be accepted at one or the other institution. Explain why James may be wrong. If possible, use a diagram in your explanation.

He may be wrong because people could get accepted at _both_ colleges. Then, others, possibly him, would not get accepted by either school.

1000 kids

100 200 300 400 500 600 700 800 900 1000

college A   college B

the poor little people with no college to go to

California Assessment Program   *Survey of Academic Skills*

OPEN-ENDED MATHEMATICS QUESTION(S)

NAME **B. D.**

Instructions: Use this sheet to answer the questions. Show as much of your work as possible. (In some cases, there may be more than one solution.) Use the reverse side of this sheet if needed.

James knows that half of the students from his school are accepted at the public university nearby. Also, half are accepted at the local private college. James thinks that this adds up to 100%, so he will surely be accepted at one or the other institution. Explain why James may be wrong. If possible, use a diagram in your explanation.

Some of the people accepted at one college could be also accepted at the other one. Therefore, some people may not be accepted to either while some people are accepted to both.

EX: 4 people → Ryan, Jeff, Josh, Steve

* Ryan and Steve get accepted to the private University. That is half the people.
* Ryan and Josh get accepted to the public college. That is half the people.
* Half + Half does not equal "everyone getting accepted" because Jeff didn't get accepted anywhere.

4

Assessing Student Growth in Mathematical Problem Solving

FRANK K. LESTER, JR., and DIANA LAMBDIN KROLL

Problem solving should be the central focus of the mathematics curriculum. (p. 23)

If problem solving is to be the focus of school mathematics, it must also be the focus of assessment... [and] assessments should determine students' ability to perform all aspects of problem solving (p. 209).

Assessment must be more than testing; it must be a continuous, dynamic, and often informal process (p. 203). — NCTM Curriculum and Evaluation Standards for School Mathematics, 1989

Although much had been written prior to the beginning of this decade about the importance of problem solving in mathematics, it seems safe to say that during the past 10 years or so problem solving has been the most written and talked about part of the mathematics curriculum. Indeed, many writers mark the beginning of the "problem-solving era" with the publication in 1980 of the *Agenda for Action* of the National Council of Teachers of Mathematics. In that document, a call was made to make problem solving "the focus of school mathematics" (NCTM, 1980, p. 1).

Since that time, a great deal of research and curriculum development has taken place concerning the upgrading of students' mathematical problem-solving abilities. Today most teachers are aware of the importance of problem solving and of the need to improve their efforts to help students become better problem solvers. As a result, rather dramatic changes have begun to take place in the K–12 mathematics curriculum. New types of problem-solving experiences are appearing in textbooks, as are new instructional techniques.

However, as promising as many of these innovations may be, there has been far too little attention paid to the related task of identifying or developing appropriate assessment techniques and instruments. Despite the appearance of reforms reflecting an emphasis on problem solving and mathematical thinking in several statewide testing programs, the extent of the changes has been far short of what is required. Moreover, teachers have been provided with almost no guidance about how to change their classroom evaluation procedures to fit these new emphases.

This chapter has a two-fold purpose: first, to present a model for the assessment of mathematical problem solving, and second, to illustrate and discuss several problem-solving assessment techniques developed over the past few years for use in mathematics classrooms.

What Is the Problem with Problem-Solving Assessment?

To set the stage for a discussion of our assessment model, we begin by posing three scenarios, each intended to illustrate a key point of this chapter.

Scenario 1

About eight years ago Frank Lester made a presentation to a group of state and district mathematics supervisors and curriculum coordinators regarding the importance of problem solving in school mathematics. At the end of the presentation, one perceptive educator insisted: "As much as I like what you're saying, I'm afraid none of your ideas have a chance of succeeding unless teachers change their evaluation methods and states change their testing programs."

This provocative statement sparked a very lively discussion about what it would take to really make problem solving the focus of school mathematics. The consensus was that curricular reform was unlikely to meet with success if it was not accompanied by a corresponding reform in the nature of mathematics assessment programs. That is, state departments of education must demonstrate a commitment to making problem solving "the central focus of the mathematics curriculum" by instituting assessment programs that are consistent with this goal.

Although the discussion at that meeting revolved around the need for those present to promote change in the assessment programs of their states, the need for reform extends to every level: state, district, school, and classroom. Just as statewide curricular change is unlikely to occur without a change in statewide assessment practices, at the level of the individual classroom, students are much more likely to view problem solving as a central part of their mathematics class if their teacher includes it as part of evaluation. Teachers who tell their students that it is important for them to show all their work but who grade papers solely on the basis of right or wrong answers are frauds, and their students soon realize it.

Scenario 2

Consider the following problem which was given to a class of fifth graders (Figure 4.1):

Eight players signed up for the tennis tournament. How many matches were played?

Figure 4.1. The tennis tournament problem.

Among the solutions arrived at by the students, those of Ali and Nathan were particularly interesting (see Figure 4.2).[1] (A modified version of this scenario appears in Charles, Lester, & O'Daffer, 1987, pp. 36–37.)

One interesting thing about their work is that Ali got an incorrect answer using an essentially correct procedure, whereas Nathan got the correct answer despite having an apparent misconception about the problem. It seems clear that Ali deserved credit for using a strategy that would have led to a correct answer had she not misunderstood or ignored one essential condition (viz., that each player played each other player exactly once). The only question is, how much credit? On the other hand, Nathan obtained a correct answer but used a procedure that is difficult to understand and may indicate a fundamental misunderstanding about the problem. In view of this concern about his work, should Nathan be given full credit? Only partial credit? No credit?

What Ali and Nathan's teacher needed was an assessment technique that takes into account considerations such as these: a technique that permits the teacher to evaluate

---

1   All names in this paper used in reference to students are pseudonymns.

(a)

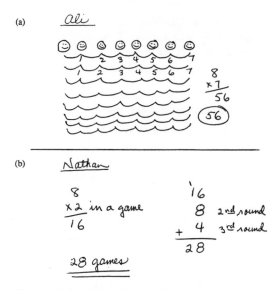

(b)

Figure 4.2. Students' written work for the tennis tournament problem.

several aspects of students' written work, not just their answers.

Scenario 3

Nathan's teacher was puzzled by what he had written on his paper. She knew that he had a tendency to write down only a very little of what he was thinking when he solved mathematics problems. Consequently, she decided that a chat was called for. Her conversation with him probably went something like this[2]:

TEACHER: Nathan, when I looked at your paper, it wasn't clear to me just how you came up with your answer. I'd like you to tell me what was going on in your head as you solved this problem. First of all, why did you write this (pointing to the computation in the upper left-hand corner of his paper)?

NATHAN: That? Oh, that's wrong! I don't know why I wrote it. I guess I should've crossed it out. (Nathan was not saying the computation was incorrectly done. Rather, he was saying that it was not appropriate for solving this particular problem.)

TEACHER: OK. Can you tell me about the rest of what you did?

NATHAN: Well, yeah. I just thought about the problem for a while and I got a picture in my head about what was going on. This (referring to the $16 + 8 + 4 = 28$ written on his paper) is just the way I kept track of the games in each round.

TEACHER: Games? Were you supposed to find out how many tennis games were played?

NATHAN: Uh, no, I mean matches.

TEACHER: All right. Tell me more about what you did.

NATHAN: OK. It's like I have a picture in my head. I paired the players off.

TEACHER: Can you draw your mental picture on this paper for me?

NATHAN: Sure! See, first of all, there are really three steps in this. First, you have this (see Figure 4.3a). I wrote an "x" for each of the eight players. Every player in the top row plays every player in the bottom row. That's 4×4 or 16 games, uh, matches. Now, the players in the top row haven't played each other and the players in the bottom row haven't played each other. So, you get 8 more matches (see Figure 4.3b). And then, you have to crisscross them to get the last matches, 4 more (see Figure 4.3c). That's 16 and 8 and 4 matches in all.

TEACHER: Nice explanation, Nathan. It would've helped me a lot if you had written some of this on your paper.

NATHAN: Yeah, I guess maybe it would've. But, I just didn't need to do that. So, I didn't.

Nathan had solved the tennis tournament problem correctly, using a very innovative approach, but there would have been no way for Nathan's teacher to figure out what Nathan had been thinking about simply by looking at his paper. A discussion of the sort shown here was essential. This example points out one limitation inherent in any analysis of students' written work—namely, that what students

---

2 The solution described by Nathan in the following dialogue was provided by Randall Charles.

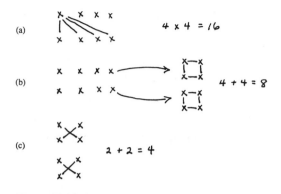

Figure 4.3. Nathan's mental picture of the tennis tournament problem.

write is often very different from what they think. Problem solving is, by its nature, a very complex type of intellectual activity. As a result, assessment of written work alone simply is not enough.

These scenarios illustrate at least three points about assessment of students' problem-solving growth: (i) assessment programs and practices should be consistent with the goals of the mathematics curriculum; (ii) teachers should use assessment techniques that take into account the processes used in solving problems as well as the answers; and (iii) analysis of written work, no matter how carefully done, cannot be the sole means of assessment.

Factors That Influence Problem-Solving Performance

By its very nature, problem solving is an extremely complex form of human endeavor that involves much more than the simple recall of facts or the application of well-learned procedures. Successful problem solving involves the process of coordinating previous experiences, knowledge, and intuition in an effort to determine an outcome of a situation for which a procedure for determining the outcome is not known.

The ability to solve mathematics problems develops slowly over a very long period of time because success depends on much more than mathematical content knowledge. Problem-solving performance seems to be a function of at least five broad, interdependent categories of factors: (i)

knowledge acquisition and utilization, (ii) control, (iii) beliefs, (iv) affects, and (v) socio-cultural contexts.

Let us say a few words about each of these categories.

Knowledge acquisition and utilization

It is safe to say that, to date, the overwhelming majority of research in mathematics education has been devoted to the study of how mathematical knowledge is acquired and utilized. Thus, it is not surprising that, traditionally, most assessment of progress in mathematics has also been directed at assessment of mathematical knowledge. Included in this category are a wide range of resources (both formal and informal) that can assist the individual's mathematical performance. Especially important types of resources are the following: facts and definitions (e.g., 12 is a composite number, a rectangle is a parallelogram with four right angles), algorithms (e.g., the regrouping algorithm for subtraction), heuristics (e.g., drawing pictures, looking for patterns, working backwards), problem schemas (i.e., packages of information about particular problem types, such as distance-rate-time problems), and the host of other routine, but nonalgorithmic, procedures that an individual can bring to bear on a mathematical task.

Of particular significance to this discussion is the realization that individuals understand, organize, represent, and ultimately utilize their knowledge in very different ways. Problem solvers must make a match between their own knowledge representation and the problem situation at hand. To the extent that they are able to achieve such a match, they are successful in solving the problem. Techniques for assessing proficiency in problem solving need to take into consideration the idiosyncratic nature both of the knowledge that individuals hold and of the ways that an individual's knowledge base impacts on his or her ability to solve problems.

Control

Even when individuals possess the mathematical knowledge and skills necessary to solve a particular problem, they are generally unsuccessful unless they are able to utilize these resources efficiently. Control refers to the marshalling and subsequent allocation of available cognitive resources to deal success-fully with mathematical situations. More specifically, it includes executive decisions about planning, evaluating, monitoring, and regulating. The processes used to regulate one's behavior are often referred to as meta-cognitive processes, and these have recently become the focus of much attention within the mathematics education research community. In fact, recent research suggests that an im-portant difference between successful and un-successful problem solvers is that successful problem solvers are much better at controlling (i.e., monitoring and regulating) their ac-tivities. A lack of control can have disastrous effects on problem-solving performance (Kroll, 1988). And there is evidence that ex-plicit attention to the metacognitive aspects of problem solving—both in instruction and in evaluation—brings the importance of monitoring behaviors to the forefront of students' awareness and can make a dif-ference in their ability to make the most of the resources and skills they have at hand (Cam-pione, Brown, & Connell, 1989).

Beliefs

Schoenfeld (1985) refers to beliefs, or "belief systems" to use his term, as the individual's mathematical world view; that is, "the perspective with which one approaches math-ematics and mathematical tasks" (p. 45). Beliefs constitute the individual's subjective knowledge about self, mathematics, the en-vironment, and the topics dealt with in par-ticular mathematical tasks. For example, many elementary school children believe not only that *all* mathematics story problems can be solved by direct application of one or more arithmetic operations, but also that the ap-propriate operation to be used is always deter-mined by "key words" in the problem (Lester & Garofalo, 1982).

Unfortunately, a student who believes that all problems can be solved by focusing on key words will probably have an incorrect solution to any problem containing misleading key words. It seems apparent that beliefs shape attitudes and emotions, and direct the decisions made during mathematical activity. In our own research we have been particularly interested in students' beliefs about the nature of problem solving, as well as about their own

capabilities and limitations (Lester, Garofalo, & Kroll, 1989b).

Affects

As is clear to any mathematics teacher who has ever admitted his or her profession in a social situation, many individuals in our society have very definite feelings related to the study of mathematics. It is not uncommon to hear people confide that they "always hated math" or "never felt confident about word problems." Clearly, the affective domain—which includes individual feelings, attitudes and emotions—is an important contributor to problem-solving behavior.

However, until quite recently, mathe-matics education research in this area was primarily limited to examinations of the cor-relation between attitudes and performance in mathematics. Not surprisingly, attitudes that have been shown to be related to performance include motivation, interest, confidence, per-severance, willingness to take risks, tolerance of ambiguity, and resistance to premature closure. In recent years, problem-solving re-searchers have become much more aware of the very pervasive nature of affective vari-ables, and subsequently much more attention has been devoted to clarifying the nature of affective variables and to identifying and studying their impact on problem-solving teaching and learning (see, for example, various chapters of McLeod & Adams, 1989)

Teachers need to be aware of the extent to which a student's performance in solving a problem may be very much influenced by af-fective factors, sometimes to the point of dominating the student's thinking and actions. Furthermore, teachers must also consider ways to include assessment of affective vari-ables in their evaluation portfolios.

Socio-cultural contexts

In recent years, the point has been raised within the cognitive psychology community that human intellectual behavior must be studied in the context in which it takes place (Brown, Collins, & Duguid, 1989; Neisser, 1976; Norman, 1981). That is to say, because human beings are immersed in a reality that both affects and is affected by human be-havior, it is essential to consider the ways in which socio-cultural factors influence cognition. In particular, it has become increasingly evident that the development, understanding, and use of mathematical ideas and techniques grow out of social and cultural situations. D'Ambrosio (1985) argues that children bring to school their own mathematics which has developed within their own socio-cultural environment. This mathematics, which he calls "ethnomathematics," provides the individual with a wealth of intuitions and informal procedures for dealing with mathematical phenomena. Furthermore, one need not look outside the school for evidence of social and cultural conditions that influence mathematical behavior. The interactions that students have among themselves and with their teachers, as well as the values and expectations that are nurtured in school, shape not only what mathematics is learned, but also how it is learned and how it is perceived (cf. Cobb, 1986). The point then is that the wealth of socio-cultural conditions that make up an individual's reality plays a prominent role in determining the individual's potential for success in doing mathematics both in and out of school. Although most socio-cultural conditions are beyond the control of teachers, awareness of them and documentation of their effects may need to be considered in designing a comprehensive problem-solving evaluation scheme for the classroom.

In actuality, the five categories discussed here overlap much more than this disjoint discussion can possibly indicate (e.g., it is clearly not possible to completely separate affects, beliefs, and socio-cultural contexts). And the categories not only overlap, but also interact in a variety of ways too numerous to name in these few pages (e.g., beliefs influence affects, and both influence knowledge utilization and control; socio-cultural contexts have an impact on all the categories).

It is perhaps due to the interdependence of these categories that problem solving is so difficult for students. Certainly, the existence of such diverse influences on problem-solving behavior makes assessment of success in mathematical problem solving much more difficult than assessment of routine mathematical knowledge and skills. On the other hand, teachers who are aware of the complex nature of these influences can much more readily understand the need for a variety of assessment techniques in evaluating the problem-solving behaviors of students. In the next section, we present a model for mathematical problem-solving assessment that includes components relating most of the five categories of factors.

A Model for Mathematical Problem-Solving Assessment

A few years ago, Frank Lester and two colleagues began to take a serious look at evaluation practices commonly used by teachers in their classrooms and by states in their testing programs (Charles, Lester, & O'Daffer, 1984, 1987). One result of their efforts was the development of a model that could be used as a guide in the development of techniques for assessing certain aspects of problem solving. This model included three components, two involving problem-solving performance and one involving features of the problems used for assessment. Two fundamental limitations of their model are that (i) it does not include a component related to affective factors or beliefs, and (ii) it does not consider control processes (i.e., monitoring progress during problem solving). In order to remedy this situation, we have revised their model by combining their two performance components, by including monitoring among the list of performance processes, and by adding a component dealing with affects and beliefs. Thus, the revised model takes into account four of the five categories of factors that we discussed in the previous section, the exception being the socio-cultural contexts category (which is a category that is not usually considered amenable to instruction, except to the extent that assessment results may suggest ways to change the culture of the classroom itself). Figure 4.4 depicts the three components of the model.

Component I: Affects and beliefs

In the preceding section, we identified *affects* as one of five categories of factors that influence problem solving. Typically, the word "affect" is used to encompass such constructs as attitudes, appreciations, preferences, emotions, and values (cf. Hart, 1989). Affects other than emotions are generally regarded as relatively long-term and stable, whereas emotions are relatively short-term and unstable

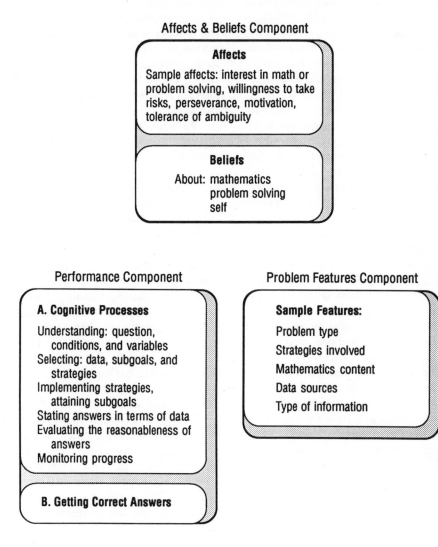

Affects & Beliefs Component

Affects

Sample affects: interest in math or problem solving, willingness to take risks, perseverance, motivation, tolerance of ambiguity

Beliefs

About: mathematics
problem solving
self

Performance Component

A. Cognitive Processes

Understanding: question, conditions, and variables
Selecting: data, subgoals, and strategies
Implementing strategies, attaining subgoals
Stating answers in terms of data
Evaluating the reasonableness of answers
Monitoring progress

B. Getting Correct Answers

Problem Features Component

Sample Features:

Problem type

Strategies involved

Mathematics content

Data sources

Type of information

Figure 4.4. A model for mathematical problem-solving assessment.

(cf. McLeod, 1989). Thus, we have included all types of affect other than emotions in our model.

As mentioned earlier, beliefs often shape attitudes and also influence the decisions made during problem solving. Beliefs that seem to play a particularly important role in the development of problem-solving ability include beliefs about the nature of mathematics and problem solving, as well as beliefs about oneself as a problem solver.

Just as instructional programs should foster the development of helpful attitudes and beliefs, assessment should endeavor to measure the extent to which they are being developed.

Component II: Performance

The performance component of the model includes two closely related subcomponents: (i) the various cognitive processes used to solve a problem, and (ii) the ability to get the correct answer. If a student is able to obtain correct answers for problems, it seems reasonable to assume that this student is also able to carry out the sorts of thinking processes involved in solving problems. However, the identification of students who can get correct answers is but one reason for assessment (and, in our minds, not the most important reason). An assessment program should be designed to provide information about the status of students' development toward becoming good problem solvers, which can then be used as a guide for subsequent instruction. Thus, even though a close relationship exists between the cognitive processes involved in problem solving and the ability to get correct answers, it is essential that an assessment model give direct attention to the range of cognitive processes that are necessary ingredients of problem-solving success. The cognitive processes subcomponent includes a number of rather broadly defined cognitive and metacognitive processes. Each of these processes is discussed briefly in the following paragraphs.

Understanding/formulating the question in a problem. One of the first tasks in solving a problem is to find or formulate the "question" and to "make sense" of it. The question is not always the last sentence in a word problem. Furthermore, it is not uncommon for the question to be posed in a statement not written in the form of a question. Making sense of the question involves understanding the meaning of specific words in the problem and also recognizing how the question relates to other statements in the problem.

Understanding the conditions and variables in the problem. Consider the following problem:

Sally gets a $5.00 allowance each week. One week her mom gave her only nickels, dimes, and quarters, a total of 24 coins in all. How many of each coin did Sally get?

The problem has two conditions: (i) the sum of the coin values is $5.00, and (ii) the total number of coins is 24. It also has three variables—the number of each kind of coin. The problem solver must recognize and consider both conditions simultaneously (as well as know the value of each type of coin) and must develop a sense of how the conditions and variables relate to each other.

Selecting the data needed to solve the problem. The problem solver must be able to identify the needed data, ignore irrelevant data, and collect and use data that may be presented in a graph, table, map, formula, or some other form. Data selection processes are closely linked to processes associated with understanding the question, conditions, and variables.

Formulating subgoals and selecting solution strategies to pursue. During the planning phase of problem solving, the problem solver must decide if there are subproblems to be solved or subgoals to be reached. Also, if relevant, decisions must be made as to the order in which these subgoals are to be pursued. A related aspect of planning is choosing a solution strategy or strategies.

Implementing the solution strategy and attaining subgoals. The problem solver must be able to both choose a strategy and implement it correctly. Implementing a strategy may involve being able to perform computations, use logical reasoning, solve equations, make an organized list of information, and so on. Similarly, after having identified and ordered a set of subgoals, the problem solver must be able to reach them.

Providing an answer in the format required in the problem. The problem solver must be able to give an answer in terms of the relevant features of the problem. It is not sufficient merely to obtain a numerical answer for a problem. The problem solver should be able to state the answer in terms of appropriate units.

Evaluating the reasonableness of an answer. The problem solver must be able to determine if an answer makes sense. This process might involve rereading the problem and checking the answer against the relevant information (question, conditions, and variables). Various estimation techniques might also be used to decide if an answer is reasonable.

Maintaining adequate control over the solution effort. Essential to successful problem solving is the ability to monitor one's thinking and actions. Effective control requires knowing *how* to monitor one's behaviors, in addition to knowing *what* and *when* to monitor.

Successful problem solving involves the process of coordinating previous experiences, knowledge, and intuition in an effort to determine an outcome of a situation for which a procedure for determining the outcome is not known. That is, successful problem solvers not only have mastered the cognitive processes of subcomponent A (and quite possibly others as well), but also can make appropriate decisions about when and how to use them. Subcomponent B of the Performance Component (getting correct answers) is included because a student who has mastered all the cognitive processes identified in subcomponent A but still cannot get correct answers would surely be considered deficient as a problem solver.

Component III: Problem features

The third component of the model specifies five important problem features that can affect a student's success in solving problems. Of course, there may well be other features that should be given attention. But whatever features the teacher decides to focus on in instruction, those features should also be varied systematically during assessment. The five features are as follows: (i) problem type, (ii) strategies that could be used to solve the problem, (iii) mathematical content and/or the types of numbers used in the problem, (iv) sources from which data need to be obtained to solve the problem, and (v) type of information included in the problem. A few words about each feature should provide an adequate description of what this component involves.

Problem type. Charles and Lester (1982) have distinguished among several types of verbal problems (viz., one-step and multiple-step translations, process problems, puzzles, and applied problems). In addition to verbal problems, two other types are pictorial/geometric and symbolic problems.

Strategies involved. The choice of problems to be included in a problem-solving assessment should be based upon a consideration of the kinds of strategies that might be useful in solving the problems. Illustrative of the sorts of strategies that might be considered in the development of an elementary or middle school assessment are the following: guess and test, work backwards, look for a pattern, use equations, use logic, draw a picture, make an organized list, make a table, act

it out, make a model, simplify, and use resources (e.g., books, calculators).

Mathematics content. As obvious as it may seem, it is important to point out that mathematics content plays an essential role in problem-solving success. A student who is able to solve a problem in one content domain may be unable to solve isomorphic versions of the problem in a different content area. For example, it is well documented that an individual who has been successful at solving problems involving whole numbers may be at a complete loss when the whole numbers are replaced by fractions or decimals (Greer, 1987). In general, the types of numbers (e.g., whole numbers, integers, fractions, decimals) and the operations on the numbers can greatly influence performance. Furthermore, mathematics content does not refer only to numbers and operations. Topics such as ratio and proportion, geometry, measurement, and algebra should also be considered.

Data sources. A written story is but one of several sources from which problem solvers must be able to extract information needed to solve problems. Problem solving often requires obtaining information from a picture, table, graph, or other source. Consequently, problem-solving assessment should also include data sources such as these.

Type of information provided in the problem statement. An important part of successful problem solving is the process of determining the information that is needed to solve the problem. In order to assess the extent to which students can identify what information is needed and can ignore what is not needed, teachers should include some problems that contain insufficient, inconsistent, or irrelevant (superfluous) information.

It would not be difficult to identify additional features of problems that should be considered in developing assessment instruments. For example, one might include syntactic features of verbal problems, or problems might be classified as to the metacognitive phases (e.g., orientation, organization, execution, verification) likely to be tapped during the solution effort (see Garofalo & Lester, 1985).

It should be pointed out that our model does not serve as a framework for assessing all aspects of problem-solving behavior. For example, it does not account for the ability to generate an "elegant" solution for a problem or the ability to identify appropriate generalizations for a solution. Rather, it points to the importance of attitudes and beliefs in problem solving, highlights several key cognitive processes that have begun to receive considerable attention in the most recently developed mathematics curricula, and it identifies several key problem features that affect problem-solving performance.

Assessment Techniques

To reiterate, the assessment model presented in the preceding section involves three major components: the first is concerned with non-cognitive aspects of students' performance (such as attitudes, preferences and beliefs), the second is related to problem-solving performance (including metacognitive behaviors), and the third involves features of the problems used for assessment. It is important for teachers to include methods for assessing each of these components in their problem-solving assessment plans. But what techniques are available for collecting such information?

This section describes a number of techniques that teachers have found useful in assessing various aspects of problem solving. These techniques are organized into four major categories: (i) observing and questioning students, (ii) using assessment data reported by students, (iii) using holistic scoring techniques, and (iv) using multiple-choice or completion tests.

Each category of techniques is discussed (with examples) in more detail in Charles et al. (1987).

Observing and questioning students

Direct observation and careful questioning of students while they solve problems can provide invaluable information not only about their performance, but also about their skills, their attitudes, and their beliefs. Direct observation and questioning is probably the best way to evaluate a student's thinking processes during problem solving. Such observation can be done either in a rather informal way, as the teacher circulates around the room, or in a more structured and formal interview.

Informal observation and questioning. A teacher can learn a lot about students' problem-solving performance and attitudes by circulating unobtrusively as they work in small groups solving problems and by interjecting questions to clarify the observations. Observation can be difficult to do, since there is so much going on when students are working on problems. When using observation as an assessment tool, it is important to consider beforehand the purpose of the assessment in order to focus attention on the relevant aspects of students' work. Teachers must ask themselves whether they are primarily interested in finding out, for example, which strategies their students understand and use, or whether they check their work regularly and efficiently, or how they feel about solving particular types of problems.

Similarly, it is important to think carefully about the types of questions one asks during observation for assessment. In everyday classroom situations, teachers ask students questions for a variety of reasons: to stimulate thinking, to provide hints, to test, or to demonstrate to other students what their peers know. Clearly, the purpose of questioning during assessment of problem solving through observation should be to help the questioner evaluate those aspects of the student's problem solving that are the focus of the observation.

Observation findings should be recorded briefly and objectively and in as timely a fashion as possible. Depending on the purpose of the observation, teachers may choose to use a comment card, a checklist, a rating scale, or a journal format for recording observations.

Structured interviews. Although direct observation and questioning are the primary components of structured interviews, structured interviews differ from informal observation in several ways. Structured interviews are generally performed with just one or two students at a time, and their structure derives from the systematic presentation of problems, probes, and questions that are planned well in advance. As in informal observation, the information obtained from a structured interview may be recorded on a rating scale, on a checklist, or in a brief narrative. But structured interviews are more likely to be audio- or videotaped for later, more detailed analysis.

Observation (either informal or structured) provides teachers with a first-hand look at how students solve problems and what they think about while solving them. And observation can be planned to assess growth in both cognitive and affective aspects of the assessment model. But when it is not feasible to talk directly with students, it may be useful to collect written self-assessment data directly from students.

Collecting written assessment data from students

Two techniques for collecting written assessment data from students are student self-reports and student self-inventories. Both of these techniques are especially appropriate for collecting information about affect, about beliefs, and about monitoring students' aspects of their problem-solving work. The usefulness of such data is dependent, of course, on how accurately, completely, and candidly students report their actions, feelings, beliefs, and intentions.

Student reports. In student self-reports, students are asked to write or to dictate into a tape recorder a retrospective account of a recently completed problem-solving experience. Usually the teacher provides some structure for the report by requiring that the student focus on questions designed to prompt commentary on selected aspects of the problem and/or the experience. Student reports can be useful, depending on how the prompting questions are designed, for assessing certain aspects of each of the three components of the problem-solving assessment model. However, they are more frequently used to assess affects or beliefs than performance or ability to handle various problem features. Student reports are certainly not appropriate for grading purposes, since such use might affect the candidness of the reports.

Student self-inventories. Another, more structured, type of student self-report is the student self-inventory. An inventory consists of a list of items, provided by the teacher, to which the student is to respond selectively (either by simply checking those items that apply, or by marking, from a range of options, the degree to which the item applies). The most familiar type of inventory is an attitude or belief survey, but inventories can also be designed to gather information about students' assessment of their own problem-solving strategies and expertise, about students' preferences for or familiarity with particular types of problems, or about students' monitoring behaviors and metacognitive awareness during problem solving. Student self-inventories can be designed to focus on any desired combination of the three components of the assessment model, since self-inventories are useful whenever student views are desired. But because such information is subjective and can be incomplete, self-inventories should probably be used only in combination with other evaluation techniques such as teacher observations or teacher-evaluated problems or test items.

Using holistic scoring techniques

Thus far, the focus of the evaluation techniques we have presented has been on students — their problem-solving skills, attitudes, beliefs, and actions — but not primarily on their written problem-solving efforts. Yet, when evaluating progress in problem solving, it makes ultimate sense to attempt to design useful ways of looking at the written work that students produce. Mathematics teachers have always done this in assessing all types of mathematical work, usually looking first at students' answers and then probing deeper for sources of errors if answers are incorrect. In problem solving, the method is similar, although more emphasis must be given to process (how the student approaches the problem) and less to product (what answer is obtained). In this section, we describe three methods for evaluating students' written work on problems: (i) analytic scoring, (ii) focused holistic scoring, and (iii) general impression scoring.

Analytic scoring. Analytic scoring involves use of a scale to assign points to certain phases of the problem-solving process. Designing an analytic scale involves first identifying the problem-solving phases that are to be evaluated, and then specifying a range of scores to be awarded for various levels of performance in each phase. An analytic scoring scale might assign four points to understanding the problem, four points to planning a solution, and two points to getting a correct answer, and include specific criteria for awarding partial credit for each phase.

Analytic scoring methods are most appropriate when it is desirable to give students feedback about their performance in key categories associated with problem solving, when it would be useful to have diagnostic information about students' specific strengths and weaknesses, or when a teacher is interested in identifying specific aspects of problem solving that may require additional instructional time. Unfortunately, use of an analytic scoring method requires considerable time to carefully analyze each students' written work.

Focused holistic scoring. Unlike analytic scoring—which produces several numeric scores, each associated with a different aspect of written problem-solving work—focused holistic scoring produces one single number assigned according to specific criteria related to the thinking processes involved in solving the problem. Focused holistic scoring is *holistic* because it focuses on the total solution as a whole (neither on the answer alone nor separately on various aspects of the solution). It is *focused* because the overall rating is a result of looking for particular, previously identified, characteristics in the students' work.

Focused holistic scoring is most appropriate when a relatively quick and superficial, yet consistent, assessment technique is needed—either as a precursor to or as a follow-up to other evaluation techniques aimed at identifying students' strengths and weaknesses in problem solving. For example, such an assessment technique might be useful for end-of-semester problem-solving examinations or for district-wide problem-solving assessment. For classroom use, focused holistic scoring should be used in combination with other, more informative, evaluation techniques.

General impression scoring. General impression scoring is an evaluation technique in which an evaluator studies a student's written problem-solving work, then relies on an overall impression to assign it a score on a scale (for example, from 0 to 4). General impression scoring is the least complicated, and the least focused, of all holistic scoring methods,

since no written criteria or rating sheets are prepared or used. Rather, the evaluator uses implicit criteria and general experience gained from examination of a wide variety of solutions to rate students' problem solutions. General impression scoring is not recommended for teachers who have had limited experience in assessing problem solving, since it is only through such experience that the perspective necessary for forming consistent general impressions is obtained.

General impression scoring is useful when it is necessary to give students feedback on a problem, and time is short. In the classroom, it might be used selectively for short assignments or quizzes, but it must be supplemented with other techniques because it does not provide the important diagnostic information necessary to modify lessons and to provide meaningful feedback to students. It is recommended that, whenever possible, teachers include written or oral feedback in addition to the single numeric evaluation that general impression scoring provides.

Using multiple choice and completion tests

Multiple choice and completion tests are clearly the least satisfactory means of assessing students' progress in mathematical problem solving. Yet, because they are expedient (easy to administer and to score), they continue to be used in classrooms and in other testing situations. Traditionally, items on such tests have focused only on whether students are able to find the correct answer to a given problem. Yet, it should be clear from the three-part assessment model we have presented that we believe there is much more to problem-solving assessment than looking for correct answers. It is, of course, possible to design multiple choice or completion items that focus on various aspects of the cognitive processes involved in problem solving rather than on correct answers alone (cf. Charles et al., 1987; Marshall, 1989; NCTM, 1989). And, through careful analysis of responses to carefully constructed items, it may be possible to diagnose student errors and misunderstandings. But, in the classroom—where other, preferable methods of evaluation are possible—we would recommend that teachers not waste their time attempting to construct and to use multiple choice or completion tests for the assessment of problem solving.

On the other hand, since multiple choice tests continue to be widely used in district-wide, state-wide, and even national assessments, it is important that teachers understand how to decide which types of items on such tests are most useful and for what types of assessment such tests are most appropriate. Multiple choice tests are better suited for measuring problem-solving performance than for measuring attitudes or beliefs. Greatly preferable to multiple choice tests that concentrate attention only on correct and incorrect answers to problems are those tests that include questions designed to tap the broad range of cognitive processes involved in problem solving (e.g., those outlined in the second component of our assessment model). Additionally, any teacher who decides to use a multiple choice test in the classroom would likely find it useful to supplement the test by allowing students to write qualifying comments about individual items. Alternatively, the teacher might provide a problem-solving inventory to go along with the test, or request that students write a self-report in which they explain some of their thought processes while solving certain of the multiple choice problems.

We have discussed a wide range of techniques for assessing students' growth in problem solving. During this discussion, we have identified some of the advantages and disadvantages of each technique, but we have not provided any sort of comparison across techniques. Tables 4.1 and 4.2 highlight our judgment as to the advantages and disadvantages of each category of techniques.

Clearly, a teacher's choice of evaluation technique needs to be based on a multitude of factors, such as the type of problem-solving skill or outcome being evaluated, the number of students being evaluated and the amount of time available, the teacher's experience in teaching and evaluating problem solving, the purpose of the evaluation, and the availability of evaluation materials. Teachers must chose techniques that both provide information appropriate for the goals of their evaluation and are feasible for use in their particular classroom situation.

How can teachers use the results of problem-solving assessment?

We believe the ultimate purpose of assessment is to aid in making instructional decisions. Perhaps, however, our organization of this chapter has put the cart before the horse by leaving the following section on how teachers can use the results of problem-solving assessment until last. All too often teachers and administrators think about assessment primarily in terms of testing in order to assign grades. We see grading as only one of four reasons — all related to instructional decision making — why teachers should assess their students' performance in problem solving.

Teachers use the results of problem-solving assessment (i) to help in communicating what is important, (ii) to make decisions about classroom climate, (iii) to make decisions about the content and methods of problem-solving instruction, and (iv) to assign grades.

Communicating what is important

As any teacher knows, students are experts at detecting what teachers consider important and unimportant. When a lesson is presented, students are bound to ask, "Do we need to know this for the test?" or "Do we have to do this, or is it optional?" If homework is assigned, but never collected, students soon catch on and stop preparing it. If the teacher claims to expect all work to be shown, but marks papers only on the basis of final answers, students soon stop writing out their work. In general, students internalize as important those aspects of instruction that their teacher emphasizes and assesses regularly. Certainly, the same is also true of instruction in problem solving. Students' attitudes and beliefs about problem solving are clearly affected by the assessment techniques that their teachers use.

For example, this section presents comments written by seventh graders who participated in a semester-long problem-solving class in which evaluation focused on many goals (both cognitive and noncognitive) other than just correct answers (Lester, Garofalo, & Kroll, 1989a). Understanding and planning were considered more important than answers, and students were expected to consider

Table 4.1. Advantages and disadvantages of different assessment techniques associated with phases of the assessment process.

Phases of assessment	Assessment techniques[a]							
	O(Inf)	O(Int)	S-Rep	S-Inv	H1	H2	H3	MC/C
Design & construction								
Time necessary to plan	Little	Medium	Little	Medium	Medium	Medium	Little	Much
Ease of design & construction	Easy	Moderate	Easy	Moderate	Moderate	Moderate	Easy	Difficult
Administering								
Can be part of everyday classwork & activities	Easily	No	Easily	Easily	As a test	As a test	As a test	As a test
Numbers of students who can be assessed at once	Few	Very few	Class	Class	Class	Class	Class	Class
Ease of administering by classroom teacher	Moderate	Difficult	Easy	Easy	Easy	Easy	Easy	Easy
Time commitment for administering	Moderate (in class)	Little (out of class)	Much (classwork)	Moderate (classwork)	Moderate (test)	Moderate (test)	Moderate (test)	Moderate (test)
Scoring & interpretation								
Time commitment for scoring & interpreting	Moderate	Much	Much	Moderate	Moderate	Moderate	Moderate	Little
Ease of scoring by classroom teacher	Moderate	Difficult	Moderate	Moderate	Moderate	Moderate	Moderate	Easy
Usefulness for assigning grades or for comparison	Little use	Almost no use	Almost no use	Useful	Useful	Useful	Useful	Useful

[a] Key to abbreviations: O(Inf) = Observation (Informal); O(Int) = Observation (Interview); S-Rep = Student self-report; S-Inv = Student self-inventory; H1 = Holistic scoring (analytic); H2 = Holistic scoring (focused); H3 = Holistic scoring (general impression); MC/C = Multiple choice/completion tests.

Table 4.2. Advantages and disadvantages of different assessment techniques associated with various special features.

Special features	Assessment techniques[a]							
	O(Inf)	O(Int)	S-Rep	S-Inv	H1	H2	H3	MC/C
Usefulness for assessing performance	Rather useful	Very useful	Little use	Little use	Useful	Useful	Useful	Somewhat useful
Usefulness for assessing affects & beliefs	Rather useful	Very useful	Rather useful	Rather useful	Little use	Little use	Little use	No use
Provides insights into student's thinking processes	Probably	Definitely	Probably	Possibly	Possibly	Possibly	Little	Little
Ease of record keeping	Difficult	Difficult	Moderate	Moderate	Easy	Easy	Easy	Easy
Potential for promoting student metacognition	High	Very high	High	High	Low	Low	Low	Very low
Potential for promoting student communication skills	Moderate	High	Very high	High	Low	Low	Low	Very low
Reliance on student's self-insight and candidness	Low	Moderate	High	High	Low	Low	Low	Low
Dependence upon clarity & completeness of student's written work	Low	Very low	Moderate	Moderate	High	High	High	Low

[a] Key to abbreviations: O(Inf) = Observation (Informal); O(Int) = Observation (Interview); S-Rep = Student self-report; S-Inv = Student self-inventory; H1 = Holistic scoring (analytic); H2 = Holistic scoring (focused); H3 = Holistic scoring (general impression); MC/C = Multiple choice/completion tests.

a range of problem-solving strategies. Time constraints were minimal. Students were expected to participate in class discussions about their problem-solving attempts, to think about their individual strengths and weaknesses as problem solvers, and to reflect back after each problem-solving experience to form their own judgment of the problem's familiarity and difficulty level. The students clearly recognized and came to value these aspects of problem solving, because they mentioned many of them six months later in essays in which they described the problem-solving class. Kathy noted that:

[The teacher] said to do the problem but not to worry about getting the correct answer. He said to concentrate on a problem solving strategy, like drawing a picture or diagram. Gradually as we became more familiar with problem solving strategies, it became more important to have a correct answer.

José wrote:

Last year's math problem solving class taught me to think more clearly. A few things that I remember are that we had a certain point scale. We got 4 points for understanding, 4 points for the work shown, and 2 points for the right answer, making a total of 10 points for each problem. After each problem we had to fill out an evaluation sheet about how we think we did on the problem and how difficult the problem was.

Todd recalled that:

[The teachers] gave each student a folder that we kept with us for our work. They would sometimes get the whole class to make a discussion about the problem(s). That would make us understand very well. To make us understand our habits and faults even better, they taped us individually or in pairs on videotape solving math problems. While you were solving the problem, you thought aloud so they could know what you were thinking. You had no certain time limit so that took a lot of pressure off.

Assessment methods communicate to students (and to parents and administrators)

what is considered important. And having an assessment method in mind also helps teachers be sure that they are including in their teaching those aspects of problem solving that they value. The assessed curriculum strongly influences what students are taught and what they value.

Making decisions about classroom climate

A classroom climate conducive to problem solving is essential to building a successful program. Among the elements that contribute to classroom climate, three are particularly important: (i) the teacher's commitment and enthusiasm—Is the teacher a problem solver too? (ii) the frequency of problem solving—Do students consider it an integral part of the class, or just an extra? and (iii) the evaluation practices used—Are students graded too often, or are they given opportunities to explore and experiment? Is persistence encouraged more than speed? Is more than the answer evaluated? Are different and innovative solutions recognized and rewarded?

Awareness of student attitudes and beliefs allows the teacher to adjust the difficulty level, variety, context, and interest of problems presented so that students are more apt to be engaged in their work. Assessment techniques such as student self-inventories, student self-reports, and interviews or observations of students can provide data necessary to make judgments about students' attitudes and beliefs about problem solving. Such data are invaluable in making decisions about the climate in the mathematics class.

Making decisions about problem-solving instruction

Assessment data can also be used to make instructional decisions about the content of a problem-solving program and the teaching methods used. Data from observations, interviews, and analyses of written work can be used to diagnose students' strengths and weaknesses. Note that the precision with which strengths and weaknesses can be diagnosed is highly influenced by the type of assessment technique used. For example, as noted earlier, student interviews provide a more detailed diagnostic profile of thought processes than holistic scoring of written work. We have found the following four areas, adapted from Polya (1957), general enough to pinpoint strengths and weaknesses reliably, yet specific enough that subsequent instruction can be prescribed: (i) understanding the problem, (ii) developing a plan, (iii) implementing the plan, and (iv) answering the question and checking the results. Ideas to consider with regard to changes in course content or teaching methods when student weaknesses are detected in these four areas are discussed in some detail in Charles et al. (1987).

Assigning grades

A final way that assessment data are used is to assign grades. It is important that teachers understand that assessment is not synonymous with grading. Every teacher should have a plan for assessing progress in problem solving, whether or not grades are assigned. When teachers decide to assign a grade, the following guidelines may be useful: (i) advise students in advance when their work will be graded; (ii) use a grading system that considers the process used to solve problems, not just the answer; (iii) be aware that pupils may not perform as well when they are to be graded; (iv) use as much assessment data, and as many different techniques, as possible as a basis for assigning grades; and (v) consider using a testing format that matches your instructional format (e.g., consider testing performance in cooperative groups if this is the way students usually work on problems).

For many students, grades are a very motivating factor. Such students will gain considerably when the system used to assign problem-solving grades reflects the many facets of problem-solving performance.

A Final Thought

In this chapter, we have described a model for the assessment of students' growth in mathematical problem solving and we have discussed a number of assessment techniques. Our intent has been to provide some much needed clarity to the current discourse about the role of assessment in mathematics instruction. Of course, if, as is recommended by the *Curriculum and Evaluation Standards for School Mathematics* (NCTM, 1989), educa-

tors decide to make a serious effort to bring assessment into better alignment with contemporary curricular emphases and instructional practices, there will be reason for real optimism about the future of mathematics instruction in our schools. However, Silver and Kilpatrick (1988) point out that, despite our best efforts, many of the most important aspects of problem-solving growth are likely to remain intractable to assessment. Appreciation of an elegant solution and willingness to take risks as a problem solver are but two examples of a host of traits that teachers should want their students to develop. However, none of the assessment techniques we have considered adequately measures these traits. Furthermore, the assessment of higher order thinking, especially thinking processes associated with problem solving, is an extremely difficult task to do well. And, even the best instruments and techniques are only as good as the person using them — namely, the teacher. If the mathematics education community expects teachers to develop the expertise needed to assess problem-solving performance, attitudes, and beliefs, it must begin now to help teachers learn not only to select and use existing assessment procedures wisely, but also to design and implement their own assessment techniques.

Acknowledgments

Portions of this chapter refer to research supported by National Science Foundation Grant No. MDR 85-50346. Any opinions, conclusions, or recommendations expressed are those of the authors and do not necessarily reflect the views of the National Science Foundation.

We are grateful to Peter Kloosterman for his helpful comments on an earlier version of this chapter

References

Brown, J. S., Collins, A., & Duguid, P. (1989). Situated cognition and the culture of learning. *Educational Researcher, 18,*(1), 32-42.

Campione, J. C., Brown, A. L., & Connell, M. L. (1989). Metacognition: On the importance of understanding what you are doing. In R. Charles & E. Silver (Eds.), *The teaching and assess-*

ing of mathematical problem solving (pp. 93-114). Reston, VA: National Council of Teachers of Mathematics.

Charles, R., & Lester, F. (1982). Teaching problem solving: What, why and how. Palo Alto, CA: Dale Seymour Publications.

Charles, R., Lester, F., & O'Daffer, P. (1984). An assessment model for mathematical problem solving. Unpublished manuscript.

Charles, R., Lester, F., & O'Daffer, P. (1987). How to evaluate progress in problem solving. Reston, VA: National Council of Teachers of Mathematics

Cobb, P. (1986). Contexts, goals, beliefs, and learning mathematics. For the Learning of Mathematics, 6, 2-9.

D'Ambrosio, U. (1985). Da realidadeàação: Reflexões sobre educação e matemática. São Paulo, Brazil: Summus Editorial.

Garofalo, J., & Lester, F. K. (1985). Metacognition, cognitive monitoring and mathematical performance. Journal for Research in Mathematics Education, 16, 163-176.

Greer, B. (1987). Non-conservation of multiplication and division involving decimals. Journal for Research in Mathematics Education, 18, 37-45.

Hart, L. E. (1989). Describing the affective domain: Saying what we mean. In D. B. McLeod & V. M. Adams (Eds.), Affect and mathematical problem solving: A new perspective (pp. 37-45). New York: Springer-Verlag.

Kroll, D. L. (1988). Cooperative mathematical problem solving and metacognition: A case study of three pairs of women. Dissertation Abstracts International, 49, 2958A (University Microfilms No. 8902580).

Lester, F. K. & Garofalo, J. (1982, April). Metacognitive aspects of elementary school students' performance on arithmetic tasks. Paper presented at the annual meeting of the American Educational Research Association, New York.

Lester, F. K., Garofalo, J., & Kroll, D. L. (1989a). The role of metacognition in mathematical problem solving. Final report to the National Science Foundation. Grant number MDR 85-50346 (available from the first author, School of Education, Indiana University, Bloomington).

Lester, F. K., Garofalo, J., & Kroll, D. L. (1989b). Self-confidence, interest, beliefs, and metacognition: Key influences on problem-solving behavior. In D. B. McLeod & V. M. Adams (Eds.), Affect and mathematical problem solving: A new perspective (pp. 75-88). New York: Springer-Verlag.

Marshall, S. P. (1989). Assessing problem solving: A short-term remedy and a long-term solution. In R. Charles & E. Silver (Eds.), Needed research on the teaching and assessment of mathematical problem solving (pp. 159-177). Reston, VA: National Council of Teachers of Mathematics.

McLeod, D. B. (1989). The role of affect in mathematical problem solving. In D. B. McLeod & V. M. Adams (Eds.), Affect and mathematical problem solving: A new perspective (pp. 20-36). New York: Springer-Verlag.

McLeod, D. B., & Adams, V. M. (Eds.). (1989). Affect and mathematical problem solving: A new perspective. New York: Springer-Verlag.

National Council of Teachers of Mathematics (1980). An agenda for action: Recommendations for school mathematics for the 1980s. Reston, VA: NCTM.

National Council of Teachers of Mathematics (1989). Curriculum and evaluation standards for school mathematics. Reston, VA: NCTM.

Neisser, U. (1976). Cognition and reality. San Francisco: W. H. Freeman.

Norman, D. A. (1981). Twelve issues for cognitive science. In D. A. Norman (Ed.), Perspectives in cognitive science (pp. 265-295). Norwood, NJ: Ablex.

Polya, G. (1957). How to solve it (2nd ed.). NY: Doubleday & Co., Inc.

Schoenfeld, A. H. (1985). Mathematical problem solving. Orlando, FL: Academic Press.

Silver, E. A., & Kilpatrick, J. (1988). Testing mathematical problem solving. In R. Charles & E. Silver (Eds.), The teaching and assessing of mathematical problem solving (pp. 178-186). Reston, VA: National Council of Teachers of Mathematics.

5

New Directions for Mathematics Assessment

GERALD KULM

The nation's attention has recently been directed to the finding that many students have a great deal of difficulty in solving the simplest mathematics problems that require thinking beyond the retrieval of practiced algorithms (Dossey et al., 1988; McKnight et al., 1987). The narrow focus on "back to basics" and the nearly unanimous decision by states and school districts in the late 1970's to settle for minimal competency in mathematics resulted in exactly the results that might have been expected. Students have learned how to do numerical computation at the expense of learning how to think and solve problems.

A byproduct of an era of competency-based education and accountability has been a focus on, if not an obsession with, testing. Falling SAT scores, poor showings in comparison with Japan and other developed countries, and failures on competency tests have resulted in even more tests at state and district levels. Tests are becoming the primary and widely accepted means for determining entry to, progress through, and exit from educational programs of all types and levels. Satisfactory performance on tests is becoming the sole indicator of the "value added" by education.

Validity of Standardized Mathematics Tests

A great deal of time, money, and effort is ex-pended by commercial testing companies to produce standardized tests that satisfy all of the requirements for reliability, norms, and validity. But how valid are these tests in measuring what really does (or ought to) go on in the nation's mathematics classrooms? Earlier this decade, debate about tests was often centered on the alignment of tests with the curriculum. Studies showed that there was a poor match between specific topics emphasized in mathematics textbooks, standardized tests, and the curriculum (Freeman et al., 1980). Results from state, national, and international tests provide another perspective on the mismatch between what students are taught and what they are tested on. A review of scores showed that students performed better on state assessments than on national and international tests in which the contents of the items may be unfamiliar (Kulm, 1986b). The Second International Mathematics Study (IMS) found, for example, that some students had the opportunity to learn only as few as 40 percent of the items in some content categories (McKnight et al., 1987).

But leaving the alignment problem aside, how well do standardized achievement tests define the nature of the mathematics that students should learn in our schools? The NCTM *Curriculum and Evaluation Standards for School Mathematics* has identified assessment standards, some of which seem difficult to attain with current approaches to standardized testing. For example, can today's standardized tests measure the ability to generate new mathematical procedures, to extend or modify familiar ones, or reflect a willingness to persevere at a mathematical task? These are mathematical abilities that are valued by the mathematics education community as important for all children to learn. Many mathematics teachers currently work hard to enable their students to achieve these kinds of learning goals. If the mathematics reform efforts that are underway have a successful impact on the way mathematics is taught, even more teachers will be attaining these goals. Given this potentially widening discrepancy between what is valued and what is tested, there appears to be a real danger that standardized achievement tests, as well as other traditionally designed tests, will underestimate children's achievement even more than they might now be doing. The judgement of the public and of policy makers is likely to continue to be based primarily on the results of these tests. The potential for underestimation of American students' mathematics achievement could have grave consequences for the continuation of a positive reform.

How well do current standardized mathematics tests reflect the extent and nature of mathematical knowledge and ability that students have? Public reports and popular media tend to focus on bottom-line results and use the most obvious interpretations. Even educators who probably know better sometimes fall into the trap of accepting at face value the overall worth, importance, or significance of achievement on standardized tests developed for their effectiveness in reflecting performance on specific, easily defined skills.

Recent reports suggest that only a small percentage of students are able to solve complex mathematics problems (Dossey et al., 1988). Research has shown, however, that many children entering first grade often know much more about numbers and addition and subtraction than previously assumed by teachers. This knowledge would not be seen through a typical paper-and-pencil test of arithmetic facts, but is revealed during interviews in which children are able to use objects and counting strategies to solve real problems (Carpenter, 1985). Similarly, in problem-solving research using one-on-one interviews, seventh and eighth graders are often able to solve algebra problems through the use of strategies such as trial and error when given sufficient time and opportunity to use non-routine methods. Many of the recent results indicating that students perform poorly in mathematics may primarily be due to the failure of the tests themselves to measure what students know and can do in mathematics. Perhaps our children are capable of performing much better than expected if given the chance to work in more open, creatively constructed testing environments.

Assessment and Teaching

An axiom of American education seems to be that "whatever get tested gets taught," or even worse, "only that which will be tested will be taught." The evidence of the past decade seems to support this view. Basic computational skills have been the focus for competency tests, spawning textbooks and instructional emphases aimed at developing these skills in students. Teachers have been legitimately concerned that if they "fight the system" and teach higher order thinking, their students would suffer on the computationally oriented tests that they are required to pass. Many educators believe that very little change will occur in mathematics curriculum and teaching without a concurrent change in testing, especially in state and national standardized tests that are used to assess and compare school-by-school achievement.

Most mathematics teachers believe that higher order thinking is important. In the Second International Mathematics Study, more than 60 percent of U.S. mathematics teachers listed their highest goal as "developing a systematic approach to solving problems and developing an awareness of the importance of mathematics to everyday life" (Crosswhite et al., 1986). Student performance on the IMS and recent National Assessment of Educational Progress (NAEP) tests indicate, however, that the aspirations of teachers and the performance of their students are very different things. Apparently, teachers are unable to accomplish what they would like to be able to do.

There are a number of plausible explanations for this contradictory state of affairs. The possibilities include (i) teachers really are trying to teach higher order thinking, but the students cannot or will not learn and/or cannot or will not apply what they learn; (ii) teachers think that they are teaching problem solving but, instead, they are teaching rote application of algorithms, narrow approaches to small classes of problems, or translation of words to equations; or (iii) although they would like to teach problem solving, teachers are not doing so because they don't know how, student competency tests don't include problem-solving items, or there isn't time. All of these explanations, along with others, are probably true to some extent for some teachers.

A perennial concern in assessment is the fear that teachers will "teach to the test," somehow subverting the more lofty notions of what should be taught. In current assessment practice, that concern is real because tests are comprised of finite, narrow categories of knowledge and skill. These skills can be mastered through memorization and practice on a fairly limited set of mathematical procedures and concepts. Further, the types of test items — usually multiple-choice — lend themselves to learning test-taking strategies and tricks that can substantially improve performance, without learning more about mathematics. The simple existence of these possibilities for subverting the purposes of assessment is an indictment of the current system.

There are some fundamental reasons for making changes in the way that mathematics learning is assessed. As alluded to earlier, valid and usable tests of higher order skills can provide an impetus for teaching higher order skills. Successful teaching of anything, including higher order thinking in mathematics, is dependent upon the ability to determine the degree to which it has been learned.

Assessment Frameworks

The traditional approach to developing assessments of mathematics uses a matrix-like framework for guiding the selection or development of test items. Often, two dimensions are used, one for the type of mathematical content and the other for cognitive processes or abilities. The latter dimension usually ranges from lower level processes (such as knowledge of facts) to higher level processes (such as application or analysis). For example, as shown in Table 5.1, the Second International Study of Mathematics (Crosswhite et al., 1986) assessment plan for 12th graders used nine mathematical topic areas and four "behavioral categories" — computation, comprehension, application, and analysis.

According to Romberg, Zarinnia, and Collis (see chapter 2, this volume), these content-by-behavior matrices, which dominate current test design, represent a old world view of mathematics learning. Both of the components of the matrices reflect outmoded and incorrect views of mathematics and of how mathematics is learned. The discrete categories of mathematical content fail to recognize the interdependence of mathematical content and do not reflect the use of "lower order" mathematical knowledge in activities such as problem solving and reasoning. The hierarchies of behavior drawn from the psychological theory of behaviorism does not account for recent work that indicates that mathematics is learned more through building complex networks of meaning, connecting new concepts and skills with those previously learned. This view is quite different from the notion of acquiring many small bits of knowledge, gradually building upon these to reach more sophisticated skills and concepts. Since assessment has such a profound effect on in-

Table 5.1. International curriculum grid.

Content topics	Behavioral categories[a]			
	I	II	III	IV
Sets and relations				
Number systems				
Algebra				
Geometry				
Elementary functions and calculus				
Probability and statistics				
Finite mathematics				
Computer science				
Logic				

[a]The cognitive levels are I, computation; II, comprehension; III, application; IV, analysis.

struction, Romberg et al., claim that curricular and instructional change in mathematics will be difficult unless these matrix models for assessment are abandoned.

A slightly different approach is used by the Council of Chief State School Officers (CCSSO), who recommended a matrix of five mathematical content areas (numbers & operations, measurement, geometry, data analysis statistics & probability, and algebra & functions) and three mathematical abilities (conceptual understanding, procedural knowledge, and problem solving) for the 1990 Fifth National Assessment of Educational Progress (CCSSO, 1988). The mathematical abilities are not intended to be hierarchical; questions in any of the three categories may be complex or relatively simple. Problem solving is seen as involving the selection of appropriate strategies and applying conceptual understanding and procedural knowledge, so that a clean division does not exist among the three categories of mathematical ability.

In an attempt to outline the task of assessing the broad range of students' mathematical knowledge and understanding, the National Council of Teachers of Mathematics Commission on Standards for School Mathematics has proposed five assessment standards that relate directly to mathematical content (NCTM, 1989) (see Table 5.2). These standards are in one sense hierarchical in that they range from mathematical knowledge to problem solving and reasoning. On the other hand, they differ from traditional taxonomies in that each of the standards includes aspects of knowledge, abilities, and applications of mathematics that would be classified by most experts as higher order thinking. That is, there is an explicit effort to avoid making sharp distinctions or to suggest that there are levels of higher and lower order thinking that define the standards topics.

In addition to mathematical content, the NCTM standards include a description of mathematical disposition, which is an important aspect of many theories of problem solving and higher order thinking. Performance on higher order thinking tasks is greatly affected by students' interest, confidence, perseverance, and ability to monitor their own progress.

Table 5.2. NCTM evaluation standards topics.

Standard 1.	Mathematical knowledge
Standard 2.	Conceptual understanding
Standard 3.	Procedural knowledge
Standard 4.	Problem solving
Standard 5.	Reasoning
Standard 6.	Mathematical disposition

Process-Oriented Outlines

Another possible framework is an outline of thinking processes typically used in solving mathematics problems. This approach guides the construction of items that focus on individual problem-solving steps or processes rather than simply on the final answer. In a pilot project to develop and field-test prototype items for assessing higher order thinking skills in science and mathematics, NAEP panelists suggested this type of framework (Blumberg et al., 1985; Kulm, 1986a). The framework consisted of six aspects of higher order thinking, drawn from a problem-solving perspective. These six aspects of higher order thinking interact in complex ways with processes that create ideas and strategies (generating) and control (evaluating/monitoring) the problem-solving event (see Table 5.3).

A similar approach was used by Schoen and Oehmke (1979) on the Iowa Problem Solving Performance (IPSP) Test, which used three problem-solving categories, based on Polya's (1957) work to guide item selection. Subtests are designed to measure skills that are assumed to be components and prerequisites of the ability to solve problems, especially verbal ones. The components are: "get to know the problem," "do it (solving the problem)," and "look back."

In Table 5.4, Charles et al. (1987) suggest seven problem-solving thinking processes as an outline to help teachers construct items to assess student performance. The chapter by Lester and Kroll in this volume describes a model for assessment that extends the work of Charles et al., and provides specific information and suggestions for applying this approach.

A more general type of framework, applied to mathematics as a special field, uses

Table 5.3. NAEP higher order skills framework.

Aspects	Processes	
	Generating	Evaluating/Monitoring
Understanding the problem or situation		
Developing a plan or investigation		
Implementing a plan or investigation		
Reaching a goal		
Assessing		
Formulating problems and goals		

information-processing abilities as an organizing theme. This approach was used, for example, by Krutetskii (1976) in a study of mathematical abilities in gifted children. A collection of 26 sets of problems were constructed and categorized within three major processes and their related subprocesses: information gathering (e.g., perception, interpretation), information processing (e.g., generalization, flexibility of thinking, reversibility of mental processes, understanding, reasoning, logic), and information retention (e.g., mathematical memory). For each of these processes, problems with arithmetic,

geometric, and algebraic content were used. In addition to information processing, Krutetskii developed items to assess the relative strength in visual-pictorial and verbal-logical types of mathematical ability.

The shifts away from traditional assessment frameworks—especially for problem solving and other forms of higher order thinking—seem to reflect attempts at a closer relationship among instruction, learning, and testing. These new frameworks also reflect some results from research in mathematics problem solving and acquisition of mathematical knowledge and skills. This research

Table 5.4. Problem-solving thinking processes and objectives.

Thinking processes	Sample objectives
1. Understand/formulate the question in the problem.	Given a problem, select, write, or state in your own words the question that will be answered when a solution is found.
2. Understand the conditions and variables in the problem.	Select/identify the key conditions and variables useful in understanding and solving the problem.
3. Select/find data needed to solve the problem.	A. Given a problem with unneeded data, identify the data needed to find a solution. B. Given a problem with missing data, find data needed for solving the problem.
4. Formulate subproblems and select an appropriate solution strategy to pursue.	A. Given a multiple-step or process problem, formulate/select subproblems that could be solved to find the solution. B. Given a problem, select a strategy that could be used to solve the problem.
5. Correctly implement the solution strategy and attain the subgoals.	Given a story problem: A. Select/draw a picture that could be used to help solve it; B. Write a number sentence that could be used to solve it.
6. Give an answer in terms of the data in the problem.	Given the numerical part of the answer to a problem, write the answer in a complete sentence.
7. Evaluate the reasonableness of the answer.	Given a problem and its answer, estimate to decide if the answer is reasonable.

suggests that assessment should be aimed at revealing the extent, complexity, and functional characteristics of mathematical thinking rather than focusing simply on final, well-formed ability or performance.

Assessment Formats and Contexts

The choice of assessment format or context seems to be based primarily on practical considerations such as administrative ease, economics, and testing criteria. These factors, all important considerations for large-scale commercial testing programs, have gradually had a great influence at the local, individual classroom level. As pointed out by the *Curriculum and Evaluation Standards* (NCTM, 1989), alignment of testing with the content and goals of mathematics instruction should not be compromised to meet psychometric or efficiency criteria. Instructional time, curricula, and teaching strategies have become focused on achieving success on standardized tests. But as pointed out earlier, these tests only marginally reflect the nature and quality of thinking processes and competencies achieved by learners who are still developing their knowledge of mathematics.

The usual context for mathematics tests, especially standardized achievement tests, is multiple choice, paper and pencil, timed, and group administered. Even classroom tests given by teachers often follow this approach, using tests supplied by the textbook company as part of an instructional package. At least three categories relating to the context in which testing occurs can be identified: time, tools, and format. A significant shift in any one of these categories has the potential for changing the nature of assessment.

The category "time" is used here to refer to the amount of time made available for a test. Since so much of school mathematics in the past has been focused on practiced skills, the completion of a large number of exercises in a fixed time period has been accepted not only as a measure of mastery, but as an indication of giftedness and potential for doing advanced work. On the other hand, higher order thinking in mathematics is by its very nature complex and multifaceted, requiring reflec-

tion, planning, and consideration of alternative strategies. Only the broadest limits on time for completion make sense on a test purporting to assess this type of thinking.

Paper and pencil are the only tools usually available to a student taking a mathematics test. But students learn mathematics using a variety of tools, including manipulatives, graph paper, scales, rulers, calculators, computers, tables of values and formulas, and textbooks. Two types of arguments are usually given for restricting these tools during testing: the tools would make the test questions trivial, or the tools would give certain students a relative advantage or disadvantage over other individuals or groups. Neither of these arguments seems sufficiently insurmountable to stand in the way of test reform. Entire school districts have, for example, made calculators available for learning and assessing mathematics (see chapter 7, this volume). The NAEP project on higher order thinking in science and mathematics demonstrated that it is logistically and technically possible, even on a national basis, to provide materials and equipment and to train test administrators to do hands-on testing with individuals and small groups (NAEP, 1987).

There are many examples available of items that exploit the use of mathematical tools in order to enrich our assessment of students' knowledge and understanding. Perhaps the simplest adaptation is to change items from the form: "Here are the data, conditions, and question. Find the answer" to "Here is the answer. Find the data and/or conditions that fit." The latter type of test item allows the use of tools, while revealing a rich range of procedural and conceptual knowledge.

The issue of relative advantage for certain students is primarily one of economics. There is little question that priorities both within and outside of education must be shifted in order to improve mathematics learning. State and local governments must support schools and communities that do not have resources such as computers and manipulatives. Even more importantly, schools must provide resources in line with these priorities, recognizing that equipment for mathematics is just as important as that for other academic and non-academic areas. Mathematics tests that allow (or require) the use of tools may be a powerful incentive for this change to take place.

The format of tests has received considerable attention by developers. The term "format" as used here refers to the item type, such as multiple choice, true-false, and so on. In mathematics, this work seems to have been aimed mainly at perfecting the characteristics of group-administered paper-and-pencil tests. Some multiple-choice tests have very creative items that are capable of assessing a surprisingly wide range and complexity of mathematical performance. On the other hand, in the words of Hunt (1986),

Thinking beings solve problems by manipulating mental models of the environment, rather than trying out responses until they find one that works. They build these models by combining their conceptualization of the problem with personal information about the world, abstracted from previous experience.

This view of problem solving suggests using more open-ended and varied formats and approaches to assessing how well people perform.

All of the recent suggestions for new assessment approaches — including student profiles and portfolios, multiple measures, case studies, learning-based approaches, and the use of technology — imply radical changes in the format and context for tests. A new view is needed in which testing is seen as a public activity with clear lines of individual and organizational responsibility for authorship, administration, and interpretation of items, tests, and indicators. Policy makers must work harder to look beyond the simple, bottom-line, numerical results of assessments. More comprehensive and thoughtful approaches should begin to replace the simplistic "report cards," group comparisons, and minimal competency emphasis of earlier accountability and indicator efforts. Progress in mathematics teaching, learning, and assessment must take place in concert with the goals and standards that are valued by the mathematics community as well as by the public, through a shared process of meaningful study and dialogue.

References

Blumberg, F., Epstein, M., & MacDonald, W. (1985). *National Assessment of Educational Progress higher order skills planning conference.* Princeton, NJ: National Assessment of Educational Progress.

Carpenter, T. P. (1985). Learning to add and subtract: An exercise in problem solving. In E. A. Silver (Ed.) *Teaching and learning mathematical problem solving: Multiple research perspectives.* Hillsdale, NJ: Lawrence Erlbaum.

Charles, R., Lester, F., & O'Daffer, P. (1987). *How to evaluate progress in problem solving.* Reston, VA: National Council of Teachers of Mathematics.

Crosswhite, F. J., Dossey, J. A., Swafford, J. O., McKnight, C. C., Cooney, T. J., Downs, F. L., Grouws, D. A., & Weinzweig, A. I. (1986). *Second International Mathematics Study, detailed report for the United States.* Champaign, IL: Stipes Publishing Company.

Council of Chief State School Officers (1988). *Assessing mathematics in 1990 by the National Assessment of Educational Progress.* Washington, DC: Author.

Dossey, J. A., Mullis, I. V., Lindquist, M. M., & Chambers, D. L. (1988). *The mathematics report card.* Princeton, NJ: Educational Testing Service.

Freeman, D. J., Kuhs, T. M., Porter, A. C., Knappen, L. B., Floden, R. E., Schmidt, W. H., & Schwille, J. R. (1980). *The fourth-grade mathematics curriculum as inferred from textbooks and tests.* Research Series No. 82, Institute for Research on Teaching. East Lansing: Michigan State University.

Hunt, E. (1986). Cognitive research and future test design. In *The redesign of testing for the 21st century.* Princeton, NJ: Educational Testing Service.

Krutetskii, V. A. (1976). *The psychology of mathematical abilities in schoolchildren.* Chicago: The University of Chicago Press.

Kulm, G. (1986a). Assessing higher order thinking in mathematics. Paper presented at the NCTM RAC/SIG Research Presession, Washington, DC.

Kulm, G. (1986b). Translating mathematics learning research into practice. In H. G. Steiner (Ed.), *Proceedings of the IDM-TME Conference on Didactics of Mathematics.* Bielefeld, West Germany: Institute fur Didaktik der Mathematik.

McKnight, C. M., Crosswhite, F. J., Dossey, J. M., Kifer, E., Swafford, J. O., Travers, K. J., & Cooney, T. J. (1987). *The underachieving cur-*

riculum: *Assessing U.S. school mathematics from an international perspective.* Champaign, IL: Stipes Publishing Company.

National Assessment of Educational Progress. (1987). *Learning by doing. A manual for teaching and assessing higher-order thinking in science and mathematics.* Princeton, NJ: National Assessment of Educational Progress.

National Council of Teachers of Mathematics. (1989). *Curriculum and evaluation standards for school mathematics.* Reston, VA: NCTM.

Polya, G. (1957). *How to solve it.* New York: Doubleday & Co.

Schoen, H. L., & Oehmke, T. M. (1979). *The IPSP problem-solving test.* Cedar Falls: University of Northern Iowa.

II
Technology and Mathematics Assessment

6

Computer-Based Assessment of Higher Order Understandings and Processes in Elementary Mathematics

RICHARD LESH

Although the central concerns of this chapter seem clear from its title — computer-based assessment of higher order understandings — each term in that title is somewhat misleading.

First, the term "higher order" suggests the incorrect notion of ideas which are "up in the air" like nebulous conceptual clouds, or on conceptual mountain tops which can only be addressed *after* "lower order" facts and skills have been mastered. But the types of understandings that will be emphasized in this chapter might better be characterized using terms such as deeper and broader. They are conceptual cornerstones which provide foundations for the most important mathematical ideas that students should learn; they are not just structurally insignificant conceptual capstones which could have been omitted if time or other instructional resources were unavailable.

The real concern of this chapter is with dimensions of understanding which are especially important, but which neglected, regardless of whether these dimensions seem to extend up, down, out, or in from traditionally emphasized knowledge. The types of deeper/higher order understandings that will be emphasized must develop on the way to learning foundation-level math concepts and principles. Otherwise, the meanings of these latter ideas will tend to be narrow, superficial, and lacking in generalizability and applicability; and they will tend to have pitifully short half-lives in memory.

Hypothesized relationships between basic facts and higher order thinking are especially important because, for real progress to be made in curriculum reform, we must avoid the excessive pendulum swings that have characterized past movements. Today's higher order objectives movement is partly a reaction to the basic skills movement of the early 1980s, which, in turn, was in large part a reaction to the "new math" movement of the 1960s and 1970s. Each of these movements had valid concerns and excesses that should be recognized and taken into account in future reform efforts. Therefore, the policy that this chapter will adopt is that a single framework of objectives must be determined which deals in a balanced and integrated way with both basic facts and skills and higher order understandings and processes; otherwise, neither emphasis is likely to succeed.

Second, the term "assessment" is often considered to refer to a unidimensional, passive indicator of the static, "high-low" state of a nonadapting organism (which may be a student, a teacher, a classroom, a school, or a school district). However, documents such as the NCTM's 1989 *Curriculum and Evaluation Standards for School Mathematics* increasingly view assessment as an integral part of complex, dynamic, self-regulating, and evolving systems in which feedbacks are used to guide adaptation. For example, when tests are used define, clarify, or monitor goals of instruction, they go beyond being neutral indicators of learning outcomes—they become powerful components of instructional treatments themselves, especially when rewards, punishments, and opportunities are based on them (as is the case in school districts from New York to Los Angeles, and from Chicago to Miami or San Antonio).

Tests are not simply passive indicators of non-adapting systems; they can have powerful positive or negative effects on curriculum improvement, depending on whether they support or subvert efforts to address desirable objectives. Therefore, the kind of computer-based tests that will be of interest here should be part of an integrated instruction-and-assessment system.

Instructional activities should inform assessment. Computer-based lessons should monitor students' on-going activities well enough so that many types of separate assessments become unnecessary. Instructional activities should themselves become primary sources of information about student progress.

Assessment activities should inform instruction. Computer-based tests should not simply be pencil-and-paper tests delivered on-line. They should involve new types of items and use questioning sequences that adapt to individual students using techniques such as computer-adaptive item selection, or that are based on an intelligent tutor or expert system. Therefore, these tests should be quicker and easier to administer, grade, and report, and they should yield more information in a given amount of time. Because multiple versions of the tests are not identical, they can be given several times during an instructional programs, and they should be able to be used to track and plan progress rather than simply to document final learning outcomes.

The preceding kinds of tests should go beyond ranking students along an unidimensional high-low scale to produce profiles of strengths and weaknesses for individual students. That is, they should also go beyond providing summative measures of final learning outcomes for students or programs to provide formative feedbacks to guide instructional decision making by students, teachers, or computer-based courseware managers. For example, they should help determine for each student, sequences of instructional activities that address priority needs and objectives while avoiding alternatives that are either too difficult or too easy at any give point in time.

Third, the term "computer-based" suggests images of an isolated student interacting with a computer terminal in a situation where social interactions are discouraged and concrete materials are not available. Nonetheless, Lesh (1990) describes a number of ways that computers can be used to facilitate and/or mediate student-to-student interactions, student-to-teacher interactions, and student-to-concrete-material interactions, and Lesh, Post, and Behr (1987) describe ways that interactive computer graphics can simulate some of the best features of desired activities for students using concrete materials or real problem-solving situations. So, by emphasizing these latter types of interactions, computer-based instruction and testing can focus on social dimensions of understanding, which are closely related to instructional goals which the NCTM's *Curriculum and Evaluation Standards* emphasizes under topic areas such as mathematics as communication, mathematics as reasoning, or mathematics as connections.

In general, mathematical ideas should develop along a variety of dimensions: for example, from external to internal (Vygotsky, 1978; Wertsch, 1980), from simple to complex (Gagne, 1985), from concrete to abstract (Piaget & Beth, 1966), and from intuitive to formal (Van Hiele, 1959). Therefore, it is important for assessment instruments to monitor progress along each of these dimensions, and it is important for items to capture some of the essential characteristics of a variety of different types of student-to-student and student-to-material interactions.

This chapter's emphasis on computer-based instruction and assessment doesn't mean that attention will be restricted to traditional types of student-to-terminal interactions, nor even to computer-related activities. In fact, the ultimate goal is to increase the effectiveness of instruction across the entire instructional setting (including homework, classwork, and computer-based activities). For example, Lesh and Lesh (1989) found that the preceding types of computer-based activities encouraged teachers to change their regular classroom activities to (i) use more concrete/familiar/graphic models, examples, and illustrations; (ii) use guided questioning techniques; (iii) focus instruction on realistic applications and problem-solving situations; (iv) focus on those teaching functions that depend most heavily on teacher-to-student interactions (e.g., coaching, mentoring, tutoring, being a role model); and (v) use technology-based problem-solving tools in both learning and problem-solving experiences. That is, students' computer-based activities can provide the basis for highly effective materials to facilitate on-the-job teacher education and staff development.

The remainder of this chapter is divided into three main sections. The first section briefly describes some of the most important principles underlying a modeling perspective of learning and assessment. This theoretical perspective is the key to the assessment strategies and definitions of objectives to be discussed. The second section describes five of the most important types of deeper-higher order objectives that should be emphasized in K–12 mathematics. The final section describes some specific types of assessment items that can be used to measure deeper-higher order understandings.

A Modeling Perspective of Learning, Instruction, and Assessment

In this chapter, the types of deeper/higher order objectives that will be identified are closely linked to (i) a MODELS-BASED view of the nature of math/science knowledge, and (ii) a MODELING view of the way new reasoning paradigms are constructed. This section will approach the topic of MODELS and MODELING from two distinct perspectives — a *mathematical perspective* and a *psychological perspective*. Using either of these

approaches, the main points will be (i) learning and problem-solving situations are interpreted by mapping them to (internal) models; (ii) early conceptualizations (or interpretations, or models) tend to be barren, distorted, and fuzzy; (iii) several alternative "correct" models may be available to interpret a given situation; and (iv) sequences of several related-but-qualitatively-distinct models may be needed to produce useful predictions in a given situation.

Therefore, (i) deeper/higher order understandings that will be identified will often have to do with students' abilities to go beyond thinking *with* particularly important conceptual models to also think *about* them — e.g., by analyzing, underlying assumptions, strengths, and weaknesses associated with each, and (ii) deeper/higher order processes that will be identified will often have to do with *translations* from one model to another, or with *transformations* of "objects" (or systems of objects) within a given model.

The examples in this section will also illustrate why, from either a mathematical or psychological perspective of modeling, the following kinds of new views of knowledge and conceptual development emerge.

Knowledge is local. In both logical and psychological domains, models have proven to be more "local" than former researchers had assumed. For example:

(i) In psychology, contrary to Piaget's notion of decalage, which was based on the assumption that tasks which are characterized by the same structure should be comparable in difficulty, mathematics/science educators have found that (in areas such as proportional reasoning) a student's ability to use a given conceptual model often varies a great deal from one situation to another, depending on a variety of contextual factors and student characteristics or beliefs (Lesh, Post, & Behr, 1988).

(ii) In mathematics, which was once considered to be about the laws of nature and about truth, discoveries such as those related to the existence of non-Euclidean geometries forced mathematicians to think of themselves as investigating the consistency (rather than the truth) of formal systems. Furthermore, when Godel and others demonstrated additional limitations of deductive systems, even consistency had to be sacrificed in favor of relative (or local) consistency. Today, mathematics is the study of structures (or models); its "objects" are structures; and to *do* mathematics is to construct, transform, and investigate the properties of these structures (or models). (For more about this point, see Steen (1987), Davis & Hersh (1981), or the AAAS's *Project 2061*.)

Knowledge exists in pieces. In any given problem solving situation, a student may have in mind a community of contradictory or partly overlapping interpretations; yet, each of these models may need to be sorted out, refined, and clarified before any of them yield useful solutions. Knowledge is not the single coherent fabric that had been imagined by earlier generations of researchers. (diSessa, 1989).

Knowledge is situated. That is, models are organized around situations and experiences as often as they are organized around abstract/conceptual categories; and "real world experience" often informs the principles on which important models are based. (Greeno, 1988; Pea, 1988).

To describe some of the most important elements of modeling, it is useful to begin with Figure 6.1, which shows a simplified version of the mathematical point of view. That is, mathematics learning and problem solving can be thought of as involving (i) a mapping from a "real world" situations into a "model world," (ii) a transformation within the "model world"

which generally produces some prediction about an event back in the modeled situation, and (iii) a prediction from the "model world" back into the "real world" situation. In realistic situations, however, modeling is more complex than Figure 6.1 suggests.

For example, as Figure 6.2 suggests, a sequence of several modeling cycles may be needed, with modifications and refinements to the model at each cycle, before the model is able to produce useful predictions in the modeled situation (i.e., before a sufficiently good fit is established between the model and the modeled situation).

As Figure 6.3 suggests, a series of models may be needed to solve a given problem. For example, the first model may consist of a schematic diagram of the problem situation, the second model may consist of a system of equations describing the most important relationships in the diagram, and the third model may consist of cartesian graphs of the preceding system of equations. That is, solutions may involve a mapping from the modeled situation to model 1, followed by a mapping from model 1 to model 2, followed by a mapping from model 2 to model 3, followed by a prediction from model 3 back into the original modeled situation.

As Figure 6.4 suggests, several models may need to be used in parallel. For example, in physics, to accurately describe the behavior of light, both a wave model and a particle model may be needed. Each model describes

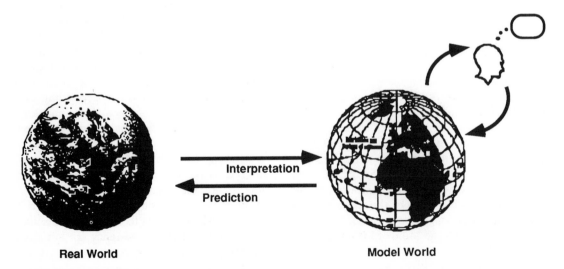

Real World **Model World**

Figure 6.1. A simplified view of mathematical modeling.

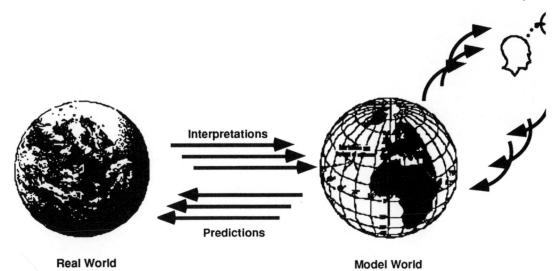

Real World

Interpretations

Predictions

Model World

Figure 6.2. Modeling may involve several cycles.

or clarifies only some aspects of the modeled situation, and each model ignores or distorts other aspects of the situation. So, a combination of all of the models may be needed to produce useful results.

Actually, realistic learning and problem solving situations often require a combination of the processes illustrated in Figures 6.2, 6.3, and 6.4. That is, in order to piece together an adequate interpretation of a reasonably complex problem solving situation, several models may need to be used in parallel and in sequence, and a series of several modeling cycles may also be needed. Furthermore, from a psychological perspective, even this composite picture of modeling is too simplistic. For in instance, examples in the remainder of this section will show that:

There are internal and external versions of each model. That is, every (external) mathematical model has an (internal) psychological counterpart which corresponds to the meanings that the model has for the human learner or problem solver.

Models tend to be unstable. That is, internal models tend to be like loosely related and shifting collections of "tectonic plates" more than like rigid and stable "worlds."

A given model tends to be embedded in several alternative notation systems (e.g., language systems, written symbol systems, diagrams) which increase their power, reduce memory load, and facilitate a variety of modeling processes (Janvier, 1987; Kaput, 1987;

Lesh, 1987). For example, (i) each notation system tends to clarify/ignore different aspects of the model, so that issues regarding to underlying assumptions must be considered; (ii) each notation system tends to have socially developed meanings which give it power and economy which is borrowed from a wide range of situations beyond the reason it is being used in any restricted problem solving situation; and (iii) each notation system tends to serves a variety of communication and recording functions which preserve or transmit thought processes across people, or within a given person across time periods, so that a variety of monitoring, checking, and reflective activities are facilitated.

In much the same way that mathematical modeling procedures may require that several external models must be used in series or in parallel, examples in the remainder of this section will show implications of the fact that several alternative representation systems, and several internal versions of these models, may also function in series or in parallel with interactions (or mappings) among them being critical for some of the most important types of higher order understandings and processes.

Interpreting learning and problem-solving situations involves mapping to internal models

One of the most important facts emerging from the forefronts of cognitive science research is that humans interpret learning and problem-solving situations by mapping them

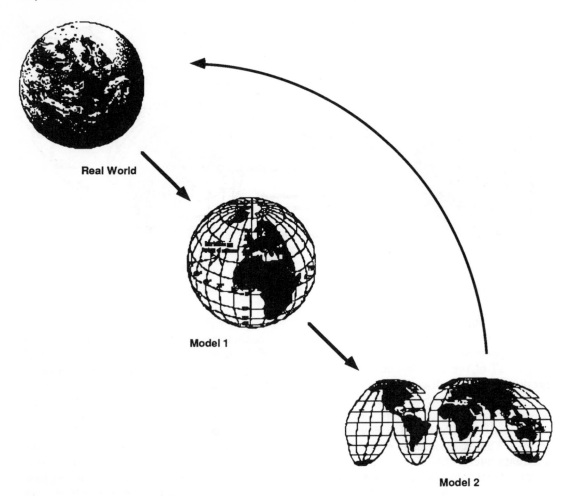

Real World

Model 1

Model 2

Figure 6.3. Modeling may involve a sequence of models.

to internal models. That is, interpretations of events are influenced by both internal model(s) and by external stimuli. For example, Figure 6.5 shows a familiar experience in which the things that one person hears doesn't match the things that are being said by another person. Because the conceptual models that are being used to interpret the given information are primitive or unstable, (i) the "big picture" is lost when attention is focused on details, (ii) only a small number of details can be kept in mind at one time (especially when attention is focused on the big picture), and (iii) some information is misinterpreted and other meanings and information are projected onto the situation which were not really present because we try to force given information to fit available models.

What kind of test could be given to assess the understandings of the pedestrian who is getting directions in Figure 6.5? One good question might be to ask for a paraphrased version (perhaps for a slightly different purpose) of the information that he has been given. Then, the quality of answers could be assessed by judging (i) How much information was noticed? (ii) Were facts and relationships perceived which were not really given? (iii) How well and how flexibly was the perceived information organized? and (iv) How sophisticated/complex/rich were the relationships that were noticed among perceived information?

According to the preceding perspective, the goal of assessment is to probe the nature of the interpreting model to determine its degree of accuracy, complexity, completeness, flexibility, and stability. But to do this, assessment must go beyond testing the amount of information noticed, and must also assess the

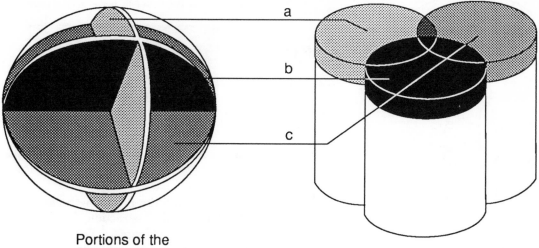

Portions of the
Real World

Figure 6.4. Modeling may involve the parallel use of several models.

nature of the patterns of information that are noticed, as well as identifying valid and invalid assumptions that are made. Similarly, the goal of instruction is not simply to get students to notice more (or different) characteristics of a given class of problem situations — the goal is to help them construct the relevant underlying model(s) which would make sense of this information once it is noticed.

Even though Figure 6.5 focuses on verbal communication, it should be clear that similar conceptual characteristics also apply to visual

and other experiences. For example, well-known facts about eye witnesses (e.g., at traffic accidents) show that different people often report seeing very different facts, and their perceptions are often amazingly barren, distorted, and internally inconsistent, depending on the sophistication and stability of their available interpreting models. Using terms from Piagetian psychology, these conceptual characteristics are referred to as (i) *centering,* which is the characteristic of noticing only the most salient features of the given situation,

Figure 6.5. Humans interpret situations by mapping to internal models.

Figure 6.6. Students' interpretations are often fuzzy, barren, and distorted.

while ignoring other relevant features, and (ii) *egocentrism,* which is the characteristic of distorting interpretations to fit prior conceptions, and consequently attributing features to the situation which aren't objectively present.

Early conceptualizations (or interpretations, or models) tend to be fuzzy and distorted

In contrast to Figure 6.1 which portrays the formal/logical view of modeling, a more accurate portrayal of the psychological view should emphasize the fact that the internal model is often a rather fuzzy and distorted version of the modeled situation.

What fraction is shaded in the picture below?

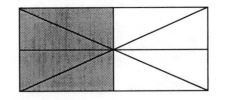

a. 4/4   b. 1/2   c. 1/1   d. not given   e. I don't know

Figure 6.7. A translation from a picture to written symbols.

To illustrate the preceding point in the context of school instruction, consider the situation shown in Figure 6.6. For the concrete models and explanations teachers present in school, students often have the same characteristics as the pedestrian in Figure 6.5; that is, their interpretations are often amazingly barren and distorted.

To illustrate the barren and distorted nature of students' interpretations of school materials, consider Figures 6.8, 6.9, 6.10, and 6.11, which show results from Lesh, Behr, and Post's research on students' understandings of rational numbers, rates, and proportional reasoning (1987). Notice that these examples focus directly on students' abilities to translate from one representation system to another. Such examples make it clear that the information students "see" and "hear" in classroom discussions (or in textbook presentations) is

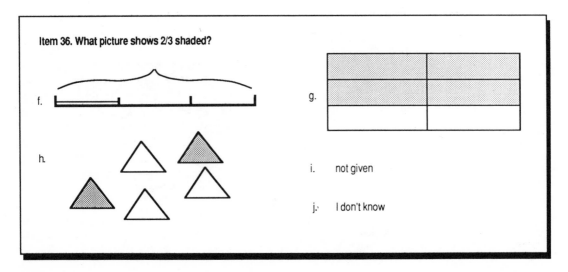

Figure 6.8. A translation from a written symbol to a picture.

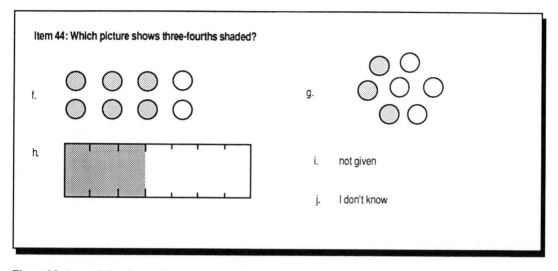

Figure 6.9. A translation from written words to a picture.

often quite different from the information that teachers (and authors) thought they were showing and describing.

In Figures 6.7 and 6.8, when attention is focused on the parts of the fraction drawings, students often lose sight of the wholes or the relevant part-whole relationships. Sixth graders answer these items correctly only 60–65 percent of the time.

In Figure 6.9, when attention is focused on wholes or on global part-whole relationships, students often fail to notice that the parts are not equivalent. Sixth graders answer this item correctly only about 20 percent of the time, with more than 40 percent selecting answer "a."

In Figure 6.9, students often limit what they see because of incorrect prior expectations, such as the expectation that three things must be compared with a nonoverlapping group of four things (rather than looking for two overlapping groups in which the ratio is three to four). Sixth graders answer this item correctly only about 37 percent of the time, with nearly 30 percent of the incorrect answers involving comparisons of three parts with four remaining parts (i.e., as in answer

choices "g" and "h").

An important point about the preceding examples is that, to improve performance on such test items, the most effective instructional techniques don't simply use the band-aid approach,[1] which identifies incorrect answers and explains correct answers to isolated test items. Instead, they tend to focus on helping students construct the underlying conceptual models that are needed to interpret the diagrams and verbal/symbolic statements (see Behr, Post, & Lesh, 1989). For example, concrete models are often used in this form of mathematics laboratory instruction, and students are encouraged to unpack and re-assemble these models using the relevant systems of relationships and operations. Then, as the operational/relational systems come to be treated as embedded parts of the concrete models, while the concrete models are used as a structural metaphors for thinking about other structurally similar problem-solving situations. So, mastery of these conceptual models is assessed using test items in which students must select, construct, unpack, or re-assemble appropriate models to deal with a variety of different types of problem-solving

---

1 The bandaid approach to instruction would be similar to trying to change the weather by lighting matches under a small number of thermometers. In much the same way that a thermometer is simply an indicator whose value can be changed without changing the weather, the preceding types of test items are indicators that tell us something about the nature of the conceptual models a student is using to interpret situations involving rational numbers, rates, and proportions. Changing students' scores on such items is not at all the same as changing the underlying model that the indicator is intended to address.

What picture shows 1/3 shaded?

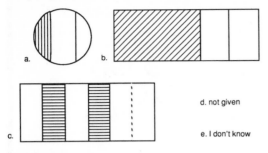

d. not given

e. I don't know

Figure 6.10. A translation from a written symbol to a picture.

situations. The stability and flexibility of these models is assessed using items that involve conceptual obstacles or perceptual distractors of the type illustrated in Figures 6.8–6.11.

Because most of the examples which have been given so far have had to do with cognitive characteristics of youngsters, it is important to emphasize these characteristics also apply to adults whenever they use (unstable) conceptual models to interpret learning and problem solving situations. That is, they are characteristics associated with modeling, not characteristics associated with people (youngsters or adults). For example, if a young child has available a stable model to interpret a given situation, then this child's cognitive characteristics would tend to seem quite adult-like. On the other hand, it is the nature of models that they *always* have some characteristics that the modeled system *does not have,* and they always *fail to have* some characteristics that the modeled system does have. If this were not the case, then the model and the modeled would be indistinguishable (i.e., one and the

same from the modeler's point of view). Therefore, to some extend, the cognitive characteristics described in this section apply whenever models are used to interpret real situations.

Several alternative "correct" models may be available to interpret a given situation

The series of pictures in Figure 6.11 can be used to show how several correct models may be possible for interpreting a given situation. For example, if only Figure 6.11d is shown prior to showing 6.11c, then most people tend to see 6.11c as being a stop sign shape or hexagon. But if shown Figures 6.11a and 6.11b prior to seeing Figure 6.11c, then most people tend to see Figure 6.11c as a cube.

Figure 6.11a can also be interpreted in more than one way. For example, the shaded side can be seen as the back side or the front side of two different cubes.

The pictures in Figure 6.11 show that the models we use to make sense of a given situation often impose relationships and organizational patterns that are not objectively given. It is often the case that information which seems to be read out of pictures (or real situations) really consists of meanings that must be read in based on models that are used to interpret the situation meaningfully. Yet, once these added meanings have been imposed, it is often difficult to assume an alternative perspective or to remember what it was like before the added meanings were recognized. For example:

(i) Once Figure 6.11c is seen as a cube, it may be difficult to re-achieve pre-conceptual innocence or to imagine what it was like

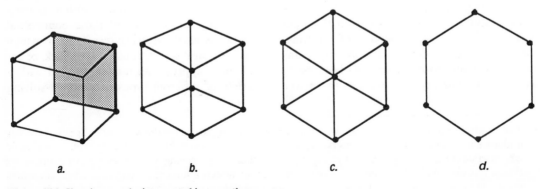

a.　　　　　　b.　　　　　　c.　　　　　　d.

Figure 6.11. Situations can be interpreted in more than one way.

before the cube was seen.

(ii) Once Figure 6.11a is seen as a cube whose front face is shaded, it may be difficult to see it as a cube slanting a different direction whose back face is shaded.

(iii) A teacher such as the one in Figure 6.6 may have difficulty understanding what it is that her students don't see in her classroom explanations and illustrations.

In some respects, the pictures in Figure 6.11 would have been more typical examples if they did not seem to emphasize *perception* more than *conception*. However, one of the most important points of this section is that a great deal of information which seems to be objectively *perceived* in real situations is in fact subjectively *conceived* by human observers. That is, (i) there may be several alternative ways to conceive (or interpret) a given figure (or verbal/symbolic statement), (ii) important characteristics of a figure (or verbal/symbolic statement) may have less to do with "raw data" than with patterns or relationships which are imposed on the data, and (iii) characteristics which are "perceived" (or ignored) depend on which organizational/relational (conceptual) frameworks are used to interpret the situation.

Because the preceding points are some of the most important for the purposes of this chapter, it is useful to illustrate them with another series of examples. Therefore, consider the following simple experiment. First, ask a third grader to view a cube from several perspectives (as indicated in Figures 6.12a and 6.12b). Then, ask the child to *"draw exactly what you see from the right-side perspective"* (i.e., the perspective corresponding to Figure 6.12b). The results often resemble Figure 6.13a or 6.13b.

Why do third graders tend to produce perspective drawings like the ones shown in Figure 6.13, whereas fourth or fifth graders tend to produce drawings that are adult-like and resemble Figure 6.11a? Inferior drawing abilities alone do not explain this transition. Reasonable explanations have more to do with the factors that make Figure 6.14 difficult for adults to draw. The fundamental difficulty is that some of the given information in Figure 6.14 doesn't make sense! It doesn't fit an available model.

In general, third graders have not yet con-

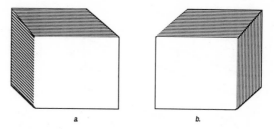

Figure 6.12. Two perspective drawings of a cube.

structed the perspective-based conceptual frameworks that older children and adults use to interpret their geometric experiences. Therefore, third graders' perspective drawings emphasize characteristics that they do understand (e.g., square angles, and equal sides) and ignore or distort characteristics that they do not understand (e.g., perspective relationships, three-dimensional connections or superpositions). Adults, on the other hand, tend to produce perspective drawings that distort the equality of the sides and the angles in order to preserve characteristics which they consider to be more important, such as characteristics associated with the perspective drawing shown in Figure 6.11a. All models oversimplify or distort some characteristics of the modeled situations in order to clarify other features.

The general significance of the preceding examples is clear if we notice that third graders' perspective drawings of a cube are very similar to instances which occurred in the history of mathematics and science. For example, even though people today tend to think it's obvious to view the world within a three-dimensional, perspective-based framework (or even a four-dimensional framework, with time as the fourth dimension), the underlying models that these frameworks presuppose

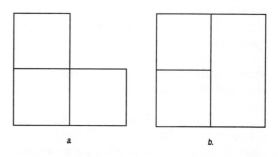

Figure 6.13. Perspective drawings by average third graders.

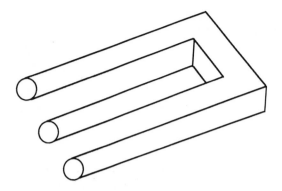

Figure 6.14. External data may not match internal models.

were not at all obvious to early civilizations. In fact, it took the genius of René Descartes to introduce the elementary-but-deep notion that rectangular coordinate systems can be imposed on the world so that equations and numbers can be used to describe whole new classes of situations, locations, or relationships. Yet, once it became natural to view the world through Descartes' three-dimensional glasses, people tended to have difficulty remembering what it was like before this conceptual framework was constructed.

Someday, in fact, when the everyday experiences of future generations are extended to include macro-events involving star galaxies and micro-events involving configurations of atoms, four-dimensional space-time dimensions may seem as obvious as three-dimensional coordinates seem today. In science and mathematics, things that are obvious tend to be very much a function of the conceptual models that we impose on reality (McCloskey, 1983).

(i) In the beginning, there were natural numbers [1,2,3,4,...]. Negative numbers were looked upon negatively, and fractions were considered to be unacceptable in polite society (as their root word actious: troublesome, quarrelsome suggests).

(ii) Later, rational numbers began to seem sensible, and integers became acceptable. But irrational numbers still didn't make any sense.

(iii) Still later, real numbers were admitted to exist. They included all integers and rational numbers, and also the formerly irrational numbers. But some things termed imaginary numbers continued to be intolerable.

(iv) Next, complex numbers—which included both real and imaginary numbers—proved to be very useful to describe important things in nature. And the struggle continues! Yet, educators continue to try to give students society's solutions to conceptual struggles that they've never recognized.

Important points to notice about the preceding examples are (i) the underlying conceptual framework(s) must be imposed on reality and cannot simply be derived from it, (ii) more than a single framework is often available to interpret a given event, (iii) different frameworks clarify and distort different given information, and (iv) once a given framework becomes stable, it is often difficult to shift to another.

From the point of view of instruction and assessment, important points to notice about the preceding examples are that (i) underlying conceptual framework(s) must be *imposed* on reality and cannot simply be derived from it; (ii) more than a single framework is often available to interpret a given event; (iii) different frameworks clarify and distort different "given" information; (iv) a set of available models may be sequentially (or hierarchically) related in the sense that one is included partly or entirely in another, and (v) once a given framework becomes stable, it is often difficult to shift to another. Therefore, conceptual shifts are as important as conceptual constructions in cognitive growth; and helping to determine when a student is "ready" to make such a shift should be one of the most important functions of an outstanding assessment program.

Items should focus on determining the extent to which important models have been constructed. To determine which model(s) a given student tends to use for an important class of problem situations, assessment questions should focus on "model discriminating items" aimed at determining which aspects are ignored and which aspects are emphasized in the situations the models are intended to describe. That is, these "model discriminating items" should also focus on (i) *the most powerful models*, which have proven to have the broadest range of applicability (i.e., models of the type the AAAS's Project 2061 is designed to identify); (ii) *the most critical questions*, which created the need for the preceding underlying models; and (iii) *the most significant conceptual discontinuities, or reorganizations, in the constructions of the preceding models.*

A sequence of several related-but-qualitatively-distinct models may be needed to produce useful predictions about a given situation

In the history of society, or in the development of a single student, conceptual growth tends to be marked by a series of discontinuities. Therefore, if an assessment program intends to address deeper and higher order issues in understanding, it should give special attention to these points of discontinuity because they tend to provide especially significant benchmarks for conceptual development. Therefore, the section that follows will give an example from science education, where a great deal of recent research has focused on clarifying the nature of students' "naive conceptualizations" of particularly important conceptual models.

Some of the most interesting recent research on related-but-qualitatively-distinct models has been done in science education investigating students' naive conceptualizations. For example, in physics education, a number of interesting naive conceptualizations have been identified which have to do with the general area of Newtonian mechanics (i.e., principles concerning forces, mass, velocity, and inertia) (Clement, 1982; diSessa, 1983, 1987; Minstrell, 1982).

Modern physics is based on the notion that the natural state of an object is to be in constant motion. Yet, everyday experience suggests that the natural state of an object is to be at rest unless some outside force has been applied. So, a number of naive conceptualizations tend to be organized around this latter perspective, and one of the goals of physics instruction is to help students recognize the need to use beyond common sense models to think about patterns and regularities beneath the surface of things.

In the case of inertia, the relevant conceptual models are based on systems of forces which are considered to remain in equilibrium. Therefore, students must construct these systems for themselves before it is reasonable to expect them to shift from old models (which seemed to explain surface characteristics of moving objects) to new and

more powerful models (which preserve deeper patterns and regularities).

To focus on deeper phenomena which lie beneath the surface of things, students must go beyond using forces to manipulate objects; they must manipulate patterns of forces, where the patterns themselves are the objects being investigated. But first, these patterns must be constructed and imposed on reality in much the same way that the coordinate systems and three-dimensional models were imposed in the preceding series of examples. If we attempt to teach students new rules (e.g., about forces and inertia) without first helping them to construct the relevant underlying models, then the understandings that result can be expected to be as barren and distorted as the pedestrian's understandings in Figure 6.5. The goal of instruction is not simply to add a few new facts or skills to students' stores of knowledge; new conceptual models must be constructed, and then students must recognize the need to make a conceptual shift from old models to new models.

(i) Because the underlying conceptual frameworks must be constructed and are not simply derived from reality, effective test items should assess the extent to which relevant constructions have occurred.

(ii) Because conceptual growth is not simply incremental and monotonic, but also involves qualitative discontinuities and digressions, it is important for assessment to help determine when students are ready for a given conceptual reorganization—or when they are ready to shift from one type of model to another.

(iii) Because a community or society of models may be available to interpret a given class of events, effective assessment and instruction activities should require students to compare and contrast alternative models in addition to focusing on the construction of individual models.

(iv) Conceptual frameworks are sorted out and refined as much as they are built up, and they also develop from concrete to abstract (Piaget & Beth, 1966), from intuitive to formal (Van Hiele, 1959), and from external to internal (Vygotsky, 1962). So, some important types of test items should be linked to benchmark capabilities along these various dimensions of development.

Knowledge in pieces, situated knowledge, and local conceptual development

Following the 1970s, when Piaget was so popular in math and science education, math/science educators traced the development of many of the most important conceptual models which should underlie the K–12 curriculum. But, whereas Piaget and his colleagues focused on concepts that developed naturally, and on periods of global conceptual reorganizations (e.g., the periods of concrete and formal operational reasoning), math/science educators tended to focus on concepts which do not develop beyond primitive levels unless artificial (mathematically rich) school learning environments are provided, and on detailed transitional stages between periods of global reorganization.

From these studies, and from related studies in cognitive science and educational anthropology, new views of conceptual development have emerged in which knowledge is not longer viewed as a "map" in which the location of a student can be pinpointed. Knowledge is not the single coherent fabric that had been imagined by earlier generations of researchers, and the old "mastery learning" perspective has proven to be hopelessly naive when it assumed that the entire mathematics/science curriculum could be broken into "bite-sized" pieces which could be mastered one at a time.

The most important principles and rules in mathematics and science tend to be linked to conceptual models which are constructed gradually, and which develop along a number of dimensions (e.g., concrete-to-abstract, intuitive-to-formal, particular-to-general), with periodic qualitative reorganizations often occurring at several points during the process of evolution. Consequently, the meaningfulness of ideas tend to increase (i) as they are embedded within progressively complex and interrelated systems of knowledge, (ii) as they are expressed to a variety of progressively powerful representation systems, and (iii) as they are linked to a variety of different problem domains and topic areas. But most importantly, the meaning of mathematics and science ideas gradually increase as the conceptual models to which they are linked become gradually refined, differentiated, and integrated into progressively complex, complete, stable, and flexible systems.

Is it a nihilistic/atomistic point of view to recognize that knowledge exists in pieces? Or that it depends on situated information? Or that it develops locally? Not at all! In fact, the opposite is true. Virtually every statement which has been made in this section about psychological knowledge also applies to mathematical models which have proven to be the most powerful and general that humans have created.

(i) Mathematical models are developed to interpret and explain local situations. However, they tend to consist of specially tailored versions of a relatively small number of particularly powerful structural metaphors.

(ii) Mathematical models are situated. That is, the principles on which they are built depend on situated experience as well as on abstract reasoning paradigms.

(iii) Mathematical models cannot always be reduced to a single piece. For example, as in the case of the wave and particle models of light, a community of contradictory or partly overlapping models may be needed to interpret and explain a given problem solving situation.

To close this section, an example will be given to illustrate some additional ways that conceptual development is similar to the process of constructing a model to interpret a learning or problem-solving situation. The example is important because it focuses attention on the *dynamics of conceptual change* — as opposed to simply focusing on the *description of conceptual states*. The example is taken from a project which focused about *Using Mathematics in Everyday Situations,* in which the most interesting types of realistic problem-solving sessions that were used were interpreted as "local conceptual development" sessions (Lesh & Kaput, 1988). That is, as students constructed models to interpret problem situations, they actually developed (locally at least) some of the most important conceptual models in elementary mathematics.

The following 40-minute problem is typical of those used in the preceding project. Such problems were designed for average-ability seventh graders who worked individually or in small groups in situations which were as much like realistic everyday experiences as we could make them in a school environment. The problem is particularly interesting because it involves a situation very similar to the one depicted earlier in this chapter in Figure 6.5; yet, it is also the kind of problem that can be adapted to a computer-based learning environment.

Near a school in a Chicago suburb, a research assistant (with a map of Chicago in hand) walks up to a group of 2–3 students and asks "What's the best[2] way to get from here to O'Hare Airport?"

When students first began to work on the O'Hare problem, alternative ways to think about the situation were seldom explicitly recognized or sorted out. Many students began by simply looking at the map to try to identify a route that was an unconscious mixture of short/quick/easy/cheap. They seldom noticed more than a few items with respect to any given way of thinking about the problem, and without explicitly being aware of it, they tended to switch back and forth among short, quick, easy, or cheap types of factors. At the same time, they often made unwarranted assumptions about the problem situation, or they imposed unnecessary constraints. For example, without examining their assumptions, they often assumed that (i) a car was available, (ii) paying for parking was not an important consideration, and (iii) leaving the car for an extended period of time presented no difficulties.

Gradually, as one or more of the preceding interpretations (short, quick, easy, cheap) became better organized and refined, more details began to be noticed, and several interpretations began to get sorted out and integrated. So, by the end of a 40-minute problem-solving session, students often generated answers which were (i) detailed (e.g., for any given route, a great many details were noticed related to factors such as time, convenience,

or possibilities of traffic jams), (ii) differentiated (e.g., different kinds of factors were sorted out and trade-offs were noted, such as the fact that potentially fast routes were often high risks for accidents or traffic jams), (iii) integrated (e.g., if you care about factors A, B, and C, then you should choose a way that deals in a combined way with these factors), (iv) conditional and flexible (e.g., if it's rush hour, then do X; if it's not, do Y or Z), and (v) aware of assumptions and possible sources of difficulties or errors (e.g., if your car is in the lot for a week, you'll pay a lot, and someone may steal your radio).

Important characteristics of the O'Hare problem include the following: (i) No tricks were needed; that is, solutions involved only straightforward uses of elementary mathematics; (ii) During 40-minute solution attempts, students seldom spent more that 5–10 minutes engaged in number crunching or answer giving activities; instead, most of their time was spent refining their interpretations of givens and goals; and (iii) Final solutions tended to involve coordinating, differentiating, and integrating several initially unstable conceptual models having to do with time, distance, cost, and so on.

Later, it will be clear why the preceding characteristics are important for assessing a variety of important types of higher order understandings. In the meantime, another characteristic that deserves special attention is that the O'Hare problem involved both "too much" and "not enough" information, with patterns, trends, and relationships in the data sometimes being as important as isolated facts. Because of this latter characteristic, models had to be constructed so that meaningful patterns could be used to (i) filter and/or simplify the situation (e.g., decisions can be based on a minimum set of cues because the model includes an explanation of how the facts are related to one another), and (ii) fill in holes, or go beyond information

---

2   In the O'Hare problem, the term "best" was left undefined because we wanted to go beyond investigating answer-giving abilities to investigate processes involved in problem formulation, information interpretation, and trial solution evaluation. Although a map was available, no other suggestions were given about whether best was intended to mean shortest, quickest, safest, simplest, least confusing, most convenient, least expensive, or some other possibility. Also, no suggestions were made about whether a car was available or whether a bus, taxi, limousine, or train might be needed. But these kinds of information were available if the students requested.

given (e.g., the model provides a holistic interpretation of the entire situation, including hypotheses about objects or events which are not present, or which needed to be actively sought out) (Bruner, 1986).

Because mathematical or psychological models tend to be useful simplifications of the things they are intended to describe, they simplify (or filter out) some aspects of reality in order to clarify (or highlight) other aspects. Yet, good models avoid oversimplifying (or overembellishing) the reality that they are intended to describe by noticing more and distorting less than competing models.

The need to resolve model-reality mismatches drives the development of models. Yet, from a psychological perspective, the ability to detect model-reality mismatches really depends upon the ability to recognize within-model mismatches. This is because external reality is only considered to have meaning insofar as it is mapped to internal models. Therefore, if the internal conceptual model is unstable (as illustrated in Figures 6.5 and 6.7 earlier in this chapter), then within-model inconsistencies may not be noticed, and consequently, model-reality mismatches may not be detected.

In general, local conceptual models evolve in the following way: (i) when a relevant system of operations and relations becomes better coordinated and more flexible, new within-model mismatches are detected; (ii) when new within-model mismatches are detected, new model-reality mismatches are also detected; and (iii) when new model-reality mismatches are detected, the need is created to develop new or improved models. These refinement cycles continue until learners or problem solvers decide that predictions from their models are good enough to achieve their goals.

In general, the preceding modeling cycles look strikingly similar regardless of whether attention is focused on the development of external mathematical models, or on the development of internal psychological models. This should not be surprising, since the refinement of internal and external models generally goes on simultaneously.

However, it is only recently that similarities started to become clear between mathematical modeling and psychological modeling. For example, until recently, mathematical models were viewed as being constructed locally to describe particular problem situations, whereas psychological models were generally considered to be Piaget-like (i.e., unitary and all encompassing) structures which de-emphasized issues having to do with the local nature of situated knowledge, the plural nature of available knowledge, the stability of systems characterized by knowledge in pieces, the importance of various representation systems, and the emphasis on modeling mechanisms for adapting existing local models to new situations. Today, however, this situation has changed! Both types of modeling are to some extent local, yet generalizable.

The points emphasized in this section have a number of important implications for teaching and testing higher order understandings and processes. For example, problem characteristics which should be emphasized include the following:

(i) A model should be constructed to deal with issues involving too much and not enough information. Since several modeling cycles will usually be required before model-reality mismatches are reduced to an acceptable level, students must go beyond thinking with these models to also think about them.

(ii) Several existing (unstable) models should be refined and integrated, so that students go beyond dealing with isolated facts and skills to create organized systems of knowledge.

(iii) The most important phases of model construction should go beyond one-step answer giving, so that a number of higher order modeling processes are emphasized. Because a series of solution steps is involved, students must plan, monitor, and check their work, in general, going beyond blind thinking to also think about thinking.

(iv) Justifying answers should often be as important as answer giving, so that the quality of answers will vary depending on the amount of information taken into account and on the extent to which assumptions, conditions, and possible error sources are identified.

As the next section of this chapter will show, some of the most important types of deeper-higher order thinking have to do with (i) going beyond thinking with conceptual models to also think about them, (ii) going beyond isolated facts and skills to organized systems of knowledge, and (iii) going beyond blind thinking to also think about thinking.

To assess the preceding types of understandings, model-construction problems have a number of advantages over more restricted types of problem solving in which students are simply required to get from (clearly identified) givens to goals when the steps are not immediately obvious. For example:

(i) Because several cycles tend to be needed to reinterpret givens and goals, answers are not simply "correct" or "not correct"; instead, the quality of correct answers can vary a great deal depending on the amounts and types of information taken into account.

(ii) In model-construction problems, the difficulty is not so much to discover a trick or a missing step as it is to execute a series of obvious steps. Therefore, even though model-construction problems tend to involve relatively complex situations, their solutions generally involve only straightforward processes.

(iii) Because model construction problems de-emphasize the importance of tricks and memorized rules, problem solving becomes an activity which is not just for the generically gifted. In O'Hare-type applied problems, we have seen that average-ability students routinely invent (or significantly extend and refine) important conceptual models which are at the heart of many of the most important deeper-higher order understandings in elementary mathematics (Lesh & Kaput, 1988).

Model-construction problems abandon the notion of assessing small and isolated facts or skills which are presumed to be either mastered or not mastered (or to exist or not exist in the student's mind). Instead, they focus on major conceptual models and aim at assessing structural attributes (e.g., stability, flexibility) along continuous scales (e.g., concrete to abstract, intuitive to formal). So, even though specific test items help to determine a student's level of mastery for a particular model, the behavioral specifications for items (e.g., "given ___, the student will ___") are

indicators of mastery; they are not definitions of mastery.

Neglected Types of Deeper-Higher Order Understandings and Processes

This section will describe five types of deeper-higher order understandings which are related to the modeling perspective described in the preceding section. From the point of view of assessment, an important observation about all five types is that deeper-higher order understandings seldom fall into the simple-minded categories of "mastered" or "not mastered." Instead, meanings gradually evolve as increases occur in the stability, flexibility, complexity, and refinement of the conceptual models and systems of relationships on which they are based. This point has important implications for assessment, because determining the existence/nonexistence of an element within a system is very different from determining the degree of stability, flexibility, complexity, or refinement of the system as a whole.

Go beyond isolated BITS of knowledge to construct well-organized SYSTEMS in which foundation-level models get special attention. Just as models enable students to filter and organize information in complex problem-solving situations, imposed systems of relationships enable students to (i) select and weigh ideas within a given topic area, and (ii) organize the domain into a stable system of knowledge. These understandings about systems of understandings are important types of higher order thinking.

(i) Concerning the selection and weighing of ideas: Within any given topic area, all ideas are not equally important. In fact, a good indicator of understanding is when students can reduce an entire chapter to notes written on a 3×5 index card. On the other hand, a typical characteristic of poor students is that every page and sentence seems equally important. So, a key component of higher order thinking involves giving special attention to the underlying conceptual models which have the greatest power and broadest range of applicability. This includes being able to (i) derive a maximum number of other ideas from this minimum set, and (ii) to use the basic models to describe a maximum number of learning and problem-solving situations.

(ii) Concerning the organization of systems of ideas: When students learn/remember/retrieve mathematics, they must go beyond learning isolated ideas and skills and must organize these understandings and processes into stable systems. Disorganized techniques and information are not knowledge any more than disorganized piles of books are a library. Also, disorganized information tends to be quickly forgotten. Part of the meaning of any given idea comes from its relationships with other ideas, and its meaningfulness and usefulness increases as it is embedded in progressively more elaborate and well-integrated systems. So, as psychologists such as Ausubel (1963) have stressed, the organization of knowledge is an important higher order objective in its own right.

As a topic area is progressively organized around a minimum sets of maximally powerful ideas, students' understandings of the entire domain increase. Furthermore, the meanings of any given idea increase each time additional ideas are linked to it and each time it is applied or extended to a new topic area. So, the meanings of deeper-higher order understandings tend to be unusually open-ended. For example:

(i) The weights or values that students assign to ideas are not entirely dictated by logic. At most, logic can endow an idea with potential power; actual power depends on the purposes, interests, and experiences of individual students.

(ii) There is no single right or wrong way to organize a system of ideas. The level of complexity and detail that is desirable may vary considerably depending on the needs and purposes of individual students. In fact, flexibility is an important characteristic of most good systems.

Organizational systems are not arbitrary. That is, certain relationships among ideas are logical necessities, and, in any topic area, there are usually some critical distinctions and generalizations—or similarities and differences—that any organizational system should recognize. On the other hand, two students should not be expected to weight and organize ideas in identical ways any more than two

textbook authors should be expected to generate identical 3×5 note cards outlining the content of those chapters. For example, in most topic areas, there are usually several alternative clusters of foundation-level ideas and derived concepts which can be selected, and the meanings of these ideas can usually be developed in a variety of ways. Or, in the minds of students, certain ideas involving arithmetic, geometry, measurement, or algebra may be thought of as being related because they have been used together in some familiar problem area—even though these ideas are were not taught in the same lessons, and even though logical relationships among them are loose.

The instruction/assessment goal is not to teach or test a single, rigid organizational system; instead, students should construct organizational systems that are stable and flexible enough to be useful for a variety of purposes and that recognize key similarities and differences within the topic area. For example, Figure 6.15 shows a computer-based note card which one student created to show the network of main ideas in his seventh grade chapter on proportional reasoning (R. Lesh & J. Lesh, 1989).

Similar electronic note cards were also created by other students in the class, with each card showing the outside layer of each student's personal data base of facts, formulas, and definitions for the chapter. Then, because the teacher allowed students to use these electronic notes on tests (or in problem-solving situations), the cards were progressively refined and simplified as they were used for a variety of purposes. The goal was for each student to gradually develop an easy-to-use cross-referencing scheme, so that information retrieval would be as quick and easy as possible for the information in their entire book. Each student was encouraged to (i) construct a networks of "big ideas" in each topic area, (ii) derive other ideas from their personal minimum set, (iii) apply and extend particularly powerful ideas in a variety of contexts, (iv) compare similarities and differences among different students' selections and organizations of big ideas, and (v) re-organize personal note cards in alternative ways for a variety of purposes.

These electronic note cards were also

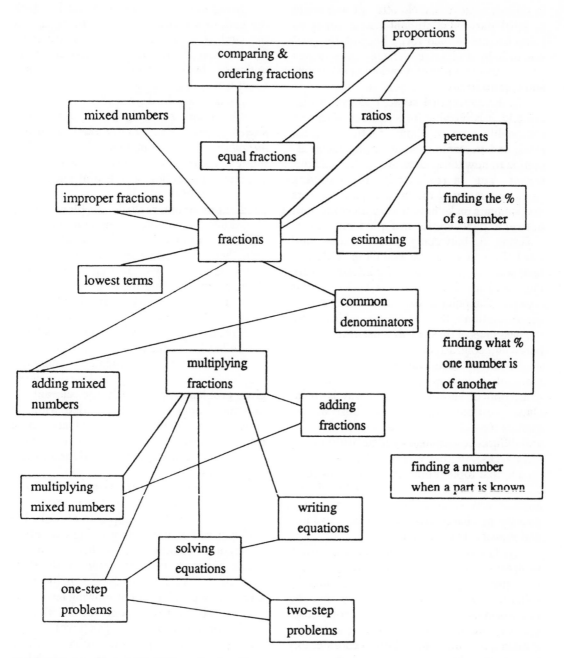

Figure 6.15. A conceptual network for rational numbers.

used by teachers on classroom-sized electronic blackboards (projected onto an overhead screen). These electronic blackboards were used for class discussions aimed at increasing the organization and flexibility of students' knowledge. In these discussion sessions, the emphasis was not on increasing the size of a body of knowledge; instead, the goal was to compare and contrast the ways different students selected and organized ideas in various topic areas. No attempt was made to label particular student's note cards as being correct or incorrect[3]. Instead, the goal was to help students understand alternatives and positive features of a variety of selections and organizations.

In the conceptual network shown in Figure 6.15, it is possible for a student to learn a reasonable amount about fractions, ratios, rates, and percents without paying much attention to how these ideas are similar and different. Yet, in realistic problem-solving situations, many of the most common errors that students make have to do with confusions between part-whole and part-part comparisons, or between fractions, ratios, and rates (Lesh, Landau, & Hamilton, 1983). In fact, some of these types of confusions were illustrated in Figures 6.8–6.11 earlier in this chapter. The point is that, in addition to dealing in isolation with rates, ratios, and fractions, the meanings of these ideas must also be compared and contrasted.

How can tests address the preceding kinds of understandings related to the organization of knowledge? One way is to include questions, such as those illustrated in Figures 6.8–6.11, which deal with similarities and differences among key ideas within main topic areas. Another way is to explicitly focus on multiple-topic problems in which concepts from a variety of areas must be combined. Still another way is to include questions aimed directly at characteristics such as flexibility and stability. For example,

(i) Questions can be included in which students must find more than one way to get from givens to goals. For example, research by cultural anthropologists has shown that the real-world math used by ordinary people (e.g., grocery shoppers, tailors, cooks, carpenters) is often quite different and often considerably more useful than the procedures they were taught in school (e.g., Carraher, Carraher, & Schliemann, 1985; Lave, 1988).

(ii) Questions can be included in which the path to a typical solution is blocked, and the student must create an alternative path. For example, Schwartz (1989) has used the notion of a broken calculator (e.g., the "3" key won't work) to get students to create alternative paths to standard types of problems.

Go beyond thinking WITH conceptual models to also think ABOUT these structural metaphors. Throughout this chapter, mathematical ideas were described as involving structural metaphors (or useful simplifications of reality) which are used to make predictions about realistic problem-solving situations. So, an important component of higher order thinking is to go beyond thinking with these conceptual models to also think about them (e.g., by evaluating their assumptions, by investigating their "goodness of fit" with the situations they are intended to describe). For example, one way to encourage students to think both with and about a foundation-level concept is ask them to use concrete materials (or manipulatable computer graphics) to act out a simplified version of the given problem. Alternatively, the relevant underlying model can be embedded within a concrete/graphic model, and students can be asked to transform, unpack, and reassemble the model in a variety of ways and for a variety of purposes.

In general, one of the best ways to focus attention on underlying models is to include problems in which one step is to describe the problem, rather than simply asking students to generate an answer or a prediction. For example, Figure 6.16 shows a trace of the steps that one student used to generate an incorrect answer to a problem about mowing lawns.

In problems like this one, the descriptions of the problem situation can involve (i) a paraphrased version of the problem; (ii) an acted-out version using concrete materials, a diagram, or a picture; or (iii) an equation or symbolic sentence. For example, using a prob-

---

3 One of the reasons why there was no single best electronic note card was that the psychological relationships that students recognized among ideas tend to be far too complex to portray in a single diagram. This situation is similar to Figure 6.12, in which two-dimensional drawings could only show selected properties of three-dimensional figures.

PROBLEM:	Bob could mow the lawn in 45 minutes, and his little brother Jim could mow it in 60 minutes. Predict how long did it take the boys when they worked together.
ESTIMATE:	About 25 minutes
PARAPHRASE:	Bob took 45 minutes per lawn. Jim took 60 minutes per lawn. How many minutes working together?
DIAGRAM:	

Bob - - - - - - - ►   Jim ──────────►

- - -|- - -|- - -|- - -|- - -|- - -|- - -|- ►

0   60   120

minutes   minutes   minutes

ARITHMETIC SENTENCE:	60 minutes/lawn + 45 minutes/lawn = [?]
STEP #1:	(ADD) 105 minutes/lawn
CHECK WITH ORIGINAL ESTIMATE:	No match?!! So the student examined the "trace" of his steps, before making a prediction.

Figure 6.15. The "trace" of solution steps to a problem.

lem-transformer utility such as the one described in Lesh, Post, and Behr (1988), students can begin with a written word problem (e.g., involving proportions, linear equations, or simple trig equations) and can give commands such as: PARAPHRASE, MODEL, SOLVE. Then, the utility can go beyond carrying out the command and can be used to judge whether the student's description is equivalent to correct prototypes.[4]

The following problem shows another way to encourage students to both think with and think about a model that is embodied in an equation or a formula. In this example, students explicitly use the model to make a prediction about a simulation of a real situation; then, they explain assumptions which might account for deviations from reality.

FACTS: It takes 20 minutes to get from Bob's house to Betty's house, and it takes 15 minutes to get from Betty's house to Ben's house.

PREDICT: How long it will take to get from Bob's house to Ben's house?

POSSIBLE FEEDBACKS (after the student has made a prediction):

(i) If the student's answer is "35 minutes," then the computer's response can be, *"My prediction is less than 35 minutes. Do you want to change your prediction before we check to see if it fits the real situation?" (Here, the computer is serving as an electronic teammate to work with the student in making predictions.)*

(ii) If the student's answer is "35 minutes," then the computer's response can be, *"That's a good prediction, but it really only takes 20 minutes. Write a brief paragraph to explain how this could happen."*

Using computer-based items, problems can include not only facts and questions, but also intelligent answer checking, helps, hints, feedbacks, follow-up questions, and information which is "available but hidden" until it is requested. Feedbacks which give "second tries" (Figure 6.15) are especially useful to encourage students to examine the models they are using and to reconsider the validity of their assumptions and procedures.

---

4   Computer-based answer checking is not restricted to numeric answers. For example, symbol manipulators can be used to check the equivalence of two numbers (e.g., $\sqrt{1/4}$ and $1/2$ and $2/4$), two expressions (e.g., $x(x + 3)$ and $x^2 + 3x$), or two functions (e.g., $a/x + b/y = 1$ and $y = (-b/a)x + b$).

Go beyond SHORT-SIGHTED THINK-ING to also THINK ABOUT THINKING. A third component of higher order understanding occurs when, during solutions to multiple-step problems, students go beyond simply thinking one step at a time to also think about their own thinking by monitoring, checking, recording, and justifying each step. Many researchers have referred to this dual-level processing as metacognition (Compione, Brown, & Connell, 1989). The aim is to help students think more intelligently by focusing on solution paths, as well as on final results. However, Vygotsky (1978) and others have shown that these capabilities are only internalized gradually and tend to evolve from external dialogues with teachers or peers. For example, the process of monitoring others eventually evolves into the ability to monitor oneself (Wertsch, 1982), and external dialogues with others eventually evolve into internal dialogues which students are able to carry on as they solve problems or construct proofs.

One way to focus on solution paths, in addition to final results, is to use multiple-concept/multiple-step problems in which (i) several answers are possible, (ii) the solution paths themselves are made explicit and can be examined, and (iii) the quality of answers depends partly on justifications that are given, including descriptions of how students arrived at their answers.

If several alternative solution paths are available, the quality of solutions often has less to do with the answers that are given than with the procedures that generated them. Furthermore, if several alternative solution paths lead to identical answers, then issues such as efficiency, simplicity, elegance, replicability, and modifiability (for future use) can be evaluated.

One advantage of using computer-based tools (e.g., spreadsheets, calculators, symbol manipulators, problem transformers) is that they not only have the ability to execute a variety of solution steps, they also have the ability to preserve solution paths, which later

can be examined, re-executed, or modified. Then, the goal of a problem may be to go beyond producing an answer to a particular question; the goal may be to produce a procedure that is capable of being replicated for a whole class of problems. Or, students can start with a solution procedure produced by someone else and debug it or make it more efficient, modify it to fit a new situation, or create an alternative sequence of steps to produce the same result.

Another technique for focusing on solution processes is to create tasks which isolate particular steps or stages in the overall solution paths. For example:

(i) Draw a picture or introduce suitable notation. The student's goal can be to construct (or assemble, or select) a picture (or diagram, or animation, or simulation) which illustrates a given problem.[5]

(ii) Look for a similar problem. The student's goal can be to construct (or write, or select, or assemble) a problem which is similar to a given problem in another context.[6]

(iii) Identify the givens and goals. If students work on messy problems in small groups (which are real or hypothetical), then this heuristic can be simplified to the form *explain (e.g., paraphrase or outline) the problem for a friend so that they can work on it too.*[7]

A simple example where several of the preceding techniques were combined uses a function machine calculator (Feurzeig, Richards, & Carter, 1988) or a word-problem assistant (Thompson, 1988), which is capable of both carrying out computations and also generating a manipulable display of the solution path. For example, if a student begins a solution by identifying relevant givens: g1, g2, g3, g4, ..., then a solution path may look like the one in Figure 6.17. Then, using the fact that this solution path is characterized by the expression $g3(g1 + g2) - g4$, a computer can check whether the solution path (as well as the final answer) are equivalent to correct prototypes.

---

5   Many graphics tools are now getting sufficiently sophisticated to be able to judge whether two constructions are equivalent.

6   Some word-problem transformer tools are now getting sufficiently sophisticated to be able to judge whether two constructions are equivalent (Lesh, Post, & Behr, 1988).

7   On a networked system of computers, students can communicate using electronic mail, which has a restricted text/graphics window in which the messages must be short and simple.

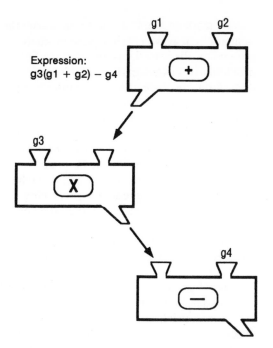

Expression:
g3(g1 + g2) − g4

Figure 6.17. A function machine solution to a problem.

Focus on realistic APPLICATIONS and PROBLEM SOLVING

No discussion of higher order reasoning would be complete without mentioning that students should get better at realistic applications and problem solving. But what are these applied problem-solving capabilities? And how are they related to the processes, skills, and heuristics that have be suggested in the past as goals for mathematics education? The following answers have been discussed throughout this chapter.

(i) Concerning givens: Realistic applied problems often involve both too much and not enough information, and patterns and trends are often as important as isolated facts. So, a model must be constructed which helps filter, organize, and fill holes in data.

(ii) Concerning goals: Realistic applied problems often have more than one correct solution, and the quality of results often depends on explanations and justifications for processes which produced them. Also, the goal(s) may consist of more than simply producing an explicitly requested mathematical result. Mathematical results may be useful tools for making predictions, decisions, comparisons, or rankings in which other factors

must be taken into account. And the quality of decisions and predictions may depend on factors such as the amount of information taken into account, the assumptions and simplifications that were made, the trends and patterns that were hypothesized, and the conditions and possible sources of errors that were identified. In real situations, goals often involve going beyond answering a particular question to create a procedure for producing solutions to a whole class of problems. Or, as Pollak (1987) has emphasized, applied problems are often of the form *"Here is a situation, describe it (in a simplified or more meaningful form)!"* rather than being of the form *"Here is a question and some data, answer it!"*

(iii) Concerning solution paths: Solutions to realistic applied problems tend to involve several modeling cycles, rather than simply involving a single given-to-goal series of steps. Furthermore, beyond-answer-giving processes tend to be particularly important, such as (i) filtering, organizing, and interpreting available information (i.e., by mapping to a model and representation system); (ii) question refinement (i.e., by making transformations in the model); and (iii) trial solution evaluation (i.e., by mapping predictions from the model back into the original problem situation).

In general, the kinds of skills that are emphasized in applied problem solving tend to go beyond number crunching to involve such things as quantifying qualitative information, making estimations and predictions, drawing informative diagrams, and writing symbolic descriptions. That is, many of the skills that are emphasized in applied problem solving involve mapping to and from various models and representation systems.

The INFLATION problem below is an example of a realistic problem that has most of the preceding characteristics. It was a problem used in Lesh's research on using math in everyday situations (Lesh & Kaput, 1988), and it generally required at least 30–40 minutes for average-ability seventh graders to solve.

MATERIALS: Students were given a calculator, a Sears catalogue from 10 years ago and a Sears catalogue today, and a newspaper from 10 years ago and a comparable newspaper today.

PROBLEM: Fred Findey began teaching here at the high school 10 years ago. He and his new bride rented an apartment at 318 Main Street for $315 per month, and he also bought a new VW Rabbit for $6,200. His starting salary was $16,300 per year. This year, Fred's brother, Tom, also began teaching at the high school, and Tom, too, just got married. In fact, he rented the very same apartment as his brother did 10 years ago, only now the rent was $600 per month, and he bought a new VW Rabbit for $13,700. Using this information, and the newspapers and catalogues, how much do you think Tom should get paid?

Even though in regular classroom teaching and testing situations, it might not often be possible to use whole books and whole newspapers as data sources (at least not with many such sources), it is not difficult to imagine how problems similar to this one could be included using computer data bases.

Go beyond DOING mathematics to also think about THE NATURE OF MATH and about THE NATURE OF PERSONAL MATHEMATICAL CAPABILITIES. A fifth important type of higher order understanding involves beliefs students form about the nature of mathematics and about their own mathematical capabilities (Schoenfeld, 1985). For example:

(i) If students' problem-solving experiences only have to do with one-rule/one-step problems which must be answered in a few seconds, they often fail to consider appropriate strategies for multiple-rule/multiple-step problem situations in which solutions are likely to take more that a few minutes to generate. Good real-world problem solvers quickly modify their strategies depending on whether a problem is judged to require a several-second solution, a several-minute solution, or a several-hour solution. So, if one-rule/one-step strategies are applied to all problems, then the solutions are often inappropriate or impossible to generate.

(ii) If students are continually confronted with overly simplistic artificial problems in which "math answers" are not necessarily the same as "real answers" (but these mismatches are never discussed, as in the prediction problems discussed earlier in this chapter), they often conclude that they must suspend their real-world knowledge in mathematics courses. For example, the following word problem was taken from a popular seventh grade textbook: *Pat could run 8 miles in an hour. How far could she run in 10 hours?*

(iii) If students never see their teachers thinking (i.e., not immediately producing the correct answers to questions), then they often form the impression that for them to be caught thinking is to be caught not knowing, which is unacceptable behavior in mathematics classes.

(iv) If geometry students think of proofs as activities in which things that are obvious are proven using strange axioms and procedures that are not very meaningful, then they are likely to develop very distorted conceptions of what geometry is all about.

The preceding kinds of misunderstandings are often referred to as attitudes, but it is useful to treat them as higher order understandings (or beliefs) about the nature of mathematics, the nature of mathematical problem solving, or the nature of personal abilities to do mathematics.

One of the main advantages that computer-based materials have over textbook-based materials is that more realistic multiple-concept/multiple-step problems can be presented because students' solution steps can be monitored and appropriate hints can be given to keep students from wasting too much time along unproductive solution paths. Also, number-crunching tools can be used so that students can work on far more problems, and on problems with realistic data, and so that they can still focus on the beyond-answer-giving phases where higher order modeling processes tend to be most important.

When computer-based lessons help provide time for teachers to observe students working on the preceding types of activities in which thought processes are easy to see, they tend to learn a lot about how students think—and about how to improve their thinking (Lesh & Lesh, 1989). Therefore, when these teachers return to their classrooms, the chances increase enormously that they will think of teaching in terms of the higher order goal of changing students' reasoning patterns rather than simply in terms of delivering facts and drilling skills.[8]

Also, when teachers are able to minimize their roles of delivery systems for low-level facts and skills, they are often able to focus their attention on instructional roles which depend most heavily on teacher-to-student interactions. For example, when computer-based instruction is used in effective ways, we have seen that teachers generally begin to refer to their emerging roles as motivators, coaches, mentors, tutors, role models, and partners in learning (Lesh & Lesh, 1989). Therefore, more attention tends to be given to the kinds of higher order beliefs described in this section.

Assessment Items and Objectives Aimed at Higher Order Thinking

This final section will focus on four types of problems which are particularly straightforward to implement on a computer, and which also emphasize some of the most important types of modeling processes emphasized throughout this chapter. These problem types are referred to as (i) interactive word problems, (ii) short stories, (iii) graph-based or table-based problems, and (iv) simulations.

The section will also describe four categories of objectives, which focus on higher order modeling processes, but which also fit easily into the "given-to-goal" format of objectives which are familiar to teachers in their textbooks and school curriculum guides. However, whereas typical behavioral objectives tend to focus on lower level facts and skills, our objectives focus on higher order processes involved in (i) analyzing and interpreting problem situations, (ii) planning and executing solution steps, (iii) monitoring and assessing intermediate or final results, and (iv) generalizing and extending solution procedures.

Interactive word problems

Computer-based word problems can be similar to textbook word problems, which are stated in only two or three sentences and which have their known and unknown quantities stated explicitly. However, the computer-based problems can be made more interesting by including (i) questions to which there is more than a single correct answer. For example, the question may involve describing the problem (e.g., using a symbolic sentence, or using graphics tools); (ii) follow-up questions which are not visible to students until responses have been given initial questions. For example, sequences of questions can be chosen from a branching set of options, where the choices at each option depend on the nature of students responses at the preceding step. So, the resulting sequence resembles a simple version of a clinical interview which might have been conducted by a human tutor. The aims include probing to determine which processes were used to generate answers, or focusing on individual solution steps (e.g., first identifying the relevant given information, then describing the problem using mathematical equations, and finally answering the question); (iii) number-crunching tools (or symbol-manipulation tools, or graphics tools) so that the student's goal is not simply to give an answer, but to use the tool to construct an answer based on givens that have been identified. Therefore, the quality of constructions can be evaluated, as well as the correctness of final results; and (iv) extra information which

---

8 In most classrooms, it is amazing how little time teachers actually have to observe individual students actively engaged in thinking, and the problem is not just that teachers don't have enough time. Pencil-and-paper activities and one-rule/one-step problems seldom provide good opportunities for teachers to observe students' reasoning patterns.

is available (perhaps at a cost in terms of the student's score on the item). For example, if some problems include "too much" and "not enough" information, then students can be given a menu of options which include (a) see additional information, (b) see a picture or diagram, (c) see a simpler (e.g., paraphrased) version of the problem, or (d) see a symbolic description of the problem.[9]

Short stories

Computer-based story problems contain far more text and information than standard word problems. For example, they may involve two or three full screens of text in which both "too much" and/or "not enough" information is given. Also, several different questions may be based on the same story. So, for each question, it is important to sort out which information is relevant from that which is irrelevant. Or, because story problems can describe more contextual factors, inferences or deductions may need to be made which are based on everyday knowledge which is not explicitly stated; or, as in the case for word problems, some missing information may need to be requested.

The goal(s) may consist of more than simply producing an explicitly requested mathematical result. Instead, students may be asked to use mathematics as a tool to make a prediction, decision, comparison, or ranking in which factors beyond logic must be taken into account.

Graph-based and/or table-based problems

Distinguishing characteristics of graph-based or table-based problems include the facts that (i) some of the relevant information may have to do with patterns or trends in the data, rather than with explicitly stated individual facts; and (ii) the graph or table may be biased in some respects, oversimplifying or distorting the underlying data that it is intended to describe. So, students may need to critically analyze the graph or table, and use patterns or trends to make reasonable predictions which go beyond (e.g., by extrapolation) or fill in holes (e.g., by interpolation) in the data.

If the goal is to make a prediction, then after a prediction has been made and the hypothetical actual result has been verified, students may be asked to identify reasonable explanations for discrepancies between the actual and predicted results. Or, in advance of seeing the actual result, students may be asked to identify possible sources or magnitudes of errors.

Simulations

Simple computer-based simulations may consist of acted-out versions of word problems or story problems. For example, instead of describing a problem situation using words, animations could be used. So, relevant information may have to be collected in somewhat the same way that students would have to collect it in a real situation.

Again, if a prediction is made, and the actual result has been shown, students may be asked to identify reasonable explanations for discrepancies between the actual and predicted results.

In the preceding types of computer-based situations, a variety of problem-solving roles can be emphasized, ranging from planning and monitoring, to data gathering and organizing, to number crunching. So, when such processes are emphasized, the following four categories of objectives can be distinguished.

Analyzing and interpreting problem situations

Analysis of the problem situation has to do with such things as identifying givens, goals, and available resources, whereas interpretation generally has to do with mapping the problem situation to some model. So, interpretation objectives often are related-use heuristics such as "draw a picture" or "look for a similar problem" (in a simpler or more familiar context).

(i) Determine the general form of the result needed to solve the problem. In a proportional reasoning problem, the general form of the solution might involve identifying the type of unknown quantity whose value was to be found (e.g., 2 miles per hour). Or, in

---

9 In Figure 6.19, when an option is chosen, a screen could appear showing a series of multiple-choice options. Then, if the student chooses the correct option (e.g., diagram, mathematical sentence, piece of relevant information), the student gets the information "free." But if an incorrect choice is made, then points are lost before the correct choice is given.

geometry, it might mean identifying the type of object or logical statement that needed to be produced.

(ii) Identify questions which can and cannot be answered (or statements which are not contradicted) by information given in the problem situation. For example, the information given in stories, tables, or graphs may be presented in biased ways (as in newspaper advertisements). Therefore, identifying supported or contradicted statements may call for assessments, deductions, or inferences beyond the information given.

(iii) Identify given information which is needed to solve a given problem. This becomes especially relevant when several questions are based on the same data source. For geometry problems, the given information is likely to consist of logical statements or attributes of given shapes. So, the values of these statements is generally true or false. For arithmetic problems, the given information tends to consist of quantities (e.g., feet, apples) whose values are numeric (e.g., 2, 1/2, $\sqrt{2}$).

(iv) Determine additional information, which is not immediately available, which is needed to solve a given problem. In some instances, this information should be available upon request, and in other cases, students may need to find a solution which is "good enough" without it (e.g., then estimate magnitudes of possible errors).

(v) Determine a similar problem (e.g., in a more familiar context). At the interpretation phase of problem solving, similarity must be judged on the basis of involving a similar pattern of relationships to the given problem. Later, at the solution generalization phase, similarity tends to be judged on the basis of being solvable using the same solution steps.

(vi) Determine a simpler problem (e.g., with values of givens modified so the solution is easier to find). In arithmetic, simpler problems often involve a special case or a particular instance. In geometry, where the givens are logical statements or attributes of objects, the values of these givens can be altered from true to false, or vice versa. This technique is sometimes called the "what-if-not" technique.

(vii) Determine a paraphrased version of the given problem. This is often done by excluding irrelevant information and/or rearranging the given information.

(viii) Determine a picture or diagram which clarifies (but which does not oversimplify or distort) the most important information in a given problem.

(ix) Determine a general description of a given situation. Often, this is done by restating the problem without giving any specific quantitative values.

(x) Determine an abstract/symbolic description of the problem situation. This often involves the heuristic "introduce suitable notation." The symbolic descriptions often involve equations, formulas, or arithmetic sentences.

(xi) Determine a subproblem toward solving a given problem. This objective has to do with partitioning the problem into simpler pieces, or working backward (or hill climbing) by identifying a series of subgoals which lead to the overall goal of the problem.

Planning and executing solution steps

Planning and executing objectives focus on paths that lead from givens to goals, rather than on the givens or goals themselves.

(i) Determine a general solution plan for getting from givens to goals without actually carrying out the steps. This might involve assembling a series of procedures from a given list of options, as though the student were writing directions for someone else to follow.

(ii) Estimate boundaries for reasonable results for a series of procedures before actually executing them. This might involve identifying possible sources of error or identifying values which are outside reasonable expectations. Such judgments are especially relevant in problems which have more than a single correct answer, or problems which involve making a predication about a real situation.

(iii) Carry out a specified series of procedures. In geometry, this may involve carrying out a given construction. However, in arithmetic, it might also be viewed as constructing solutions or following directions.

(iv) If unanticipated barriers to progress appear during a solution attempt, identify alternative solution paths.

Monitoring and assessing solution intermediate or final results

Monitoring and assessing objectives involve

being a sort of "bird on your own shoulder"; at the same time you are doing things, you can also watch yourself doing them. That is, they involve metacognition, or thinking about thinking. However, they also have to do with flexibility in adapting to unanticipated opportunities or barriers to planned progress. So, students not only monitor and assess what they are doing, they must also monitor and assess changing conditions in the problem situation.

(i) Observe someone else executing a correct series of procedures, and identify an execution error.

(ii) Identify unnecessary or inappropriate steps in a given solution plan.

(iii) Identify new opportunities, or new information, which might appear during the course of a solution attempt.

(iv) After a model has been used to make a prediction about a real situation, and after the real situation has been observed, identify factors which explain any differences between predicted and actual results.

Generalizing and extending overall solution procedures

Generalizing and extending objectives have to do with the model that was used to interpret the problem situation, and the procedures that were used to solve the problem. In general, generalizing and extending objectives have to do with the preparation of models and procedures so that they will be useful in future situations.

(i) After solving a given problem, identify other similar problems in which the same solution procedure could be applied.

(ii) After solving a given problem, identify other, more complex problems in which the existing solution procedure might be used as a significant substep.

(iii) Simplify an inefficient solution procedure.

(iv) After solving a given problem using one procedure, determine an alternative procedure.

(v) After solving a given problem, determine how changes in the conditions of the problem would have effected the results. One of the goals here has to do with reversibility of thought. For example, if conditions A, B, C, ..., N produced result R, then what changes in the conditions would be needed to produce result R'?

(vi) After solving a given problem using one set of procedures, analyze the quality of the procedure (perhaps comparing it with another procedure) using criteria such as the amount of information taken into account, the reasonableness of assumptions, the possible sources of error, the efficiency of procedures, and so on.

The preceding objectives fit with the modeling perspective which has been emphasized throughout this chapter. Yet, they also fit with traditional problem-solving objectives of the type that are often identified in standardized tests, state and district school curriculum guides, and popular textbooks. Furthermore, they are stated in a way that capitalizes on the capabilities of computer-based assessment and instruction, and in such a way that fits with the future-oriented recommendations of professional organizations (e.g., the NCTM's *Curriculum and Evaluation Standards for School Mathematics* (1989)).

The goal of this chapter has been to show how a modeling perspective of mathematics can be used to clarify the nature of some of the most important deeper/higher order objectives that educators should address, and to describe some specific types of items and evaluation criteria which are straightforward to implement in computer-based testing environments. In some respects, the proposals have been modest, because they can be implemented NOW—using existing technologies and software capabilities. On the other hand, they imply a level of integration which does not yet exist among instruction, assessment, implementation, and technological innovation. Therefore, the goal has been to try to provide some of the theoretical scaffolding which will make this integration possible.

References

American Association for the Advancement of Science (1989). *Project 2061: What science is most worth knowing?* Washington, DC.

Ausubel, D. (1963). *The psychology of meaningful verbal learning.* New York: Grune & Stratton.

Bauersfeld, H. (1979). Hidden dimensions in the so-called reality of a mathematics classroom. In R. Lesh & W. Secada (Eds.), *Some theoretical issues in mathematics education: Papers from a research presession.* Columbus, OH: ERIC Clearinghouse.

Bishop, A. (1985). The social psychology of mathematics education. In L. Streefland (Ed.), *Ninth international conference on the psychology of mathematics learning.* Utrecht, Netherlands: University of Utrecht Press.

Bruner, J. (1973). In J. Anglin (Ed.), *Beyond the information given.* New York: Norton & Co.

Bruner, J. (1986). *Actual minds, possible worlds.* Cambridge, MA: Harvard University Press.

Campione, J., Brown, A., & Connell, M. (1989) Metacognition: on the importance of understanding what you are doing. In R. Charles & E. Silver (Eds.), *The teaching and assessing of mathematical problem solving.* Hillsdale, NJ: Lawrence Erlbaum Associates.

Carraher, T., Carraher, D., & Schliemann, A. (1985). Mathematics in the streets and the schools. *British Journal of Devleopmental Psychology, 3,* 21-29.

Clement, J. (1982). Students' preconceptions in introductory mechanics. *Americal Journal of Physics, 50,* 66-71.

diSessa, A. (1982). Unlearning Aristotelian physics: a study of knowledge-based learning. *Cognitive Science,* vol. 6, no. 1, 37-75.

diSessa, A. (1983). Phenomenology and the evolution of intuition. In D. Gentner & A. Stevens (Eds.), *Mental Models.* Hillsdale, NJ: Lawrence Erlbaum Associates.

diSessa, A. (1987). Toward an epistemology of physics. Cognitive Science Technical Report. Berkeley, CA: Institute for Cognitive Science.

diSessa, A. (1989) Knowledge in pieces. In G. Gorman & P. Pufall (Eds.), *Constructivism in the Computer Age.* Hillsdale, NJ: Lawrence Erlbaum Associates.

Dienes, Z. (1960). *Building up mathematics.* London: Hutchinson Educational Ltd.

Dweck, C., & Leggett, J., (1988). A social-cognitive approach to motivation and personality. *Psychological Review, 95,* 256-273.

Feurzeig, W., Richards, J., & Carter, R. (1988). *Visdom: a visual programming language* (computer program). Sunnydale, NY: Sunburst Communications.

Gagné, R. (1985). *The conditions of learning and theory of instruction,* 4th edition. New York: Holt, Rinehart & Winston.

Greeno, J. (1988). The situated activities of learning and knowing mathematics. In M. Behr, C. Lacampagne, & M. Wheeler (Eds.), *Proceedings of the tenth annual meeting of the Psychology of Mathematics Education.* DeKalb, IL.

Janvier, C. (1987). *Problems of representation in teaching and learning mathematics.* Hillsdale, NJ: Lawrence Erlbaum Associates.

Kaput, J. (1989). Linking representations in the symbol systems of algebra. In C. Kieran & S. Wagner. *Research issues in the learning and teaching of algebra.* Hillsdale, NJ: Lawrence Erlbaum Associates.

Kelly, B. (1955). *The psychology of personal constructs.* New York: Norton.

Lancy, D. (1983). *Cross-cultural studies in cognition and mathematics.* New York: Academic Press.

Lampert, M. (1988). Connecting mathematical teaching and learning. In E. Fennema, T. Carpenter, & S. Lamon (Eds.), *Integrating research on teaching and learning mathematics.* Madison, WI: Wisconsin Center for Education Research.

Lampert, M. (1988). The teacher's role in reinventing the meaning of mathematical knowing in the classroom. In M. Behr, C. Lacampagne, & M. Wheeler (Eds.), *Proceedings of the tenth annual meeting of PME-NA.* DeKalb, IL.

Lave, J. (1988). *Cognition in practice.* New York: Cambridge University Press.

Lave, J., Smith, S., & Butler, M. (1989). Problem solving as an everyday activity. In R. Charles & E. Silver (Eds.), *Teaching and measuring problem solving.* Reston, VA: National Council of Teachers of Mathematics.

Lesh, R. (1987). The evolution of problem representations in the presence of powerful conceptual amplifiers. In C. Janvier (Ed.), *Problems of representation in teaching and learning mathematics.* Hillsdale, NJ: Lawrence Erlbaum Associates.

Lesh, R. (1990). Computer-based instruction and social dimensions of understanding. *Educational Studies in Mathematics.*

Lesh, R., Behr, M., & Post, T. (1987). Rational number relations and proportions. In C. Janvier (Ed.), *Problems of representation in teaching and learning mathematics.* Hillsdale, NJ: Lawrence Erlbaum Associates.

Lesh, R., & Kaput, J. (1988). Interpreting modeling as local conceptual development. In J. DeLange & M. Doorman (Eds.), *Senior secondary mathematics education.* Utrecht, Netherlands: OW&OC.

Lesh, R., Landau, M., & Hamilton, E. (1983). Conceptual models in applied mathematical problem solving research. In R. Lesh & M. Landau (Eds.), *Acquisition of mathematics concepts and processes* (pp. 263-343). New York: Academic Press.

Lesh, R., & Lesh, J. (1989). On-the-job teacher education. In I. Wirszup & R. Streit (Eds.), *Developments in school mathematics education around the world.* Reston, VA: National Council of Teachers of Mathematics.

Lesh, R., Post, T., & Behr, M. (1987). Dienes revisited: Multiple embodiments in computer environments. In I. Wirszup & R. Streit (Eds.), *Developments in school mathematics education around the world.* Reston, VA: National Council of Teachers of Mathematics.

Lesh, R., Post, T., & Behr, M. (1988). Proportional reasoning. In M. Behr & J. Hiebert (Eds.), *Number concepts & operations in the middle grades.* Reston, VA: National Council of Teachers of Mathematics.

Lesh, R., & Zawojewski, J. (1987). Problem solving. In T. Post (Ed.), *Teaching mathematics in grades K–8: Research-based methods.* Boston: Allyn & Bacon.

McCloskey, M. (1983). Intuitive physics. *Scientific American, 284,* 222.

McLeod, D. (1988). Research on learning and instruction in mathematics: The role of affect. In E. Fennema, T. Carpenter, & S. Lamon (Eds.), *Integrating research on teaching and learning mathematics.* Madison, WI: Wisconsin Center for Education Research.

Minstrell, J. (1982). Conceptual development research in the natural setting of a secondary school science classroom. In M. B. Rowe (Ed.), *Education for the 80's: Science.* Washington, DC: National Education Association.

Noddings, N. (1985). Small groups as a setting for research on mathematical problem solving. In E. Silver (Ed.), *Teaching and learning mathematical problem solving: Multiple research perspectives,* pp. 345-360. New York: Lawrence Erlbaum Associates.

Piaget, J., & Beth, E. (1966). *Mathematical epistemology and psychology.* Dordrecht, Netherlands: D. Reidel.

Pollak, H. (1987). Notes from a talk given at the Mathematical Sciences Education Board. Frameworks Conference, May 1987, at Minneapolis.

Polya, G. (1957). *How to solve it.* Princeton, NJ: Princeton University Press.

Richards, J., Feurzeig, W., & Carter, R. (1988). *The algebra workbench* (computer program). Sunnydale, NY: Sunburst Communications.

Rogoff, B., & Lave, J. (Eds.) (1984). *Everyday cognition: Its development in social context.* Cambridge, MA: Harvard University Press.

Schoenfeld, A. (1985). *Mathematical problem solving.* Orlando, FL: Academic Press.

Schwartz, J. (1989). Intensive quantity and referent transforming arithmetic operations. In J. Hiebert & M. Behr (Eds.), *Number Concepts and Operations in the Middle Grades.* Hillsdale, NJ: Lawrence Erlbaum Associates.

Steen, L.A. (1988). The science of patterns. *Science, 240,* 611-616.

Thompson, P. (1987). *Expressions* (computer program). Normal: Illinois State University, Department of Mathematical Sciences.

Thompson, P. (1988). *Word problem assistant* (computer program). Normal: Illinois State University, Department of Mathematical Sciences.

van Hiele, P.M. (1959). La pens'ee de l'enfant it la geometrie. *Bulletin de l'Association des Professeurs Mathematiques de l'Enseignement Public, 198,* 199-205.

Vygotsky, L. (1962). *Thought and language.* Cambridge, MA: MIT Press.

Vygotsky, L. (1978). *Mind in society: The development of the higher psychological processes.* Cambridge, MA: Harvard University Press.

Webb, N. (1982). Group composition, group interaction and achievement in cooperative small groups. *Journal of Educational Psychology, 74* (4), 475-484.

Wertsch, J. (1980). The adult-child dyad as a problem solving system. *Child Development, 51,* 1215-1221.

Wilder, R. (1981). *Mathematics as a cultural system.* New York: Pergamon.

Wirszup, I., & Streit, R. (Eds.) (1987, 1989). *Developments in school mathematics education around the world.* Reston, VA: National Council of Teachers of Mathematics.

7

Calculators and Mathematics Assessment[1]

DOROTHY STRONG

The National Council of Teachers of Mathematics (NCTM), in its position paper, *Calculators in the Mathematics Classroom*, called for the use of calculators in assessment. According to NCTM,

The evaluation of student understanding of mathematical concepts and their application, including standardized tests, should be designed to allow the use of the calculator.

The paper further recommended that authors and test writers integrate the use of the calculator into their mathematics materials at all grade levels (NCTM, 1986).

When the Chicago public schools decided to purchase calculators for all students, the anxiety of educators, parents, and society in general regarding the effect of the calculator on computational skills was apparent. Many people expressed concerns that students would no longer master basic computational skills.

Mathematics educators have warned about the danger of total abandonment of paper-and-pencil skills. On the other end of the spectrum were teachers who feared that teaching with a calculator available would be unfair if calculators were used to support instruction but students were not allowed to use

them in testing situations. Realizing the importance of the support of teachers and the community, the Chicago school system was determined that a successful program must include two important components: (i) assurance of mastery and maintenance of computational skills, and (ii) calculators as an integral part of the testing program. This solution, though trivial, proved to be profound. A dual testing technique appears to have satisfied most of the concerns regarding the use of calculators for mathematics instruction.

The tests are organized into two parts: Part I, the noncalculator section, evaluates mastery of the objectives for arithmetic computations. This component of the test assesses students' mastery and maintenance of estimation, mental computations, and paper-and-pencil computational skills. Students are expected to demonstrate 80 percent mastery on the 10 items included in this section of the tests, which they must do without calculator support.

Part II, the calculator-supported component of the tests, evaluates students' mastery of the objectives for mathematics concepts, strategies, processes, and applications. Mastery of mathematical skills in seven areas of learning—arithmetic, quantitative relationships, measurement, algebraic concepts,

---

1 All figures are copyright 1987 by the Board of Education of the City of Chicago.

geometry, probability and statistics, and applications—is assessed. At each grade level, four of the seven areas are identified each quarter for instruction and assessment. Students are expected to demonstrate mastery at 80 percent level in each of the four areas identified. Students may use calculators with this part of the test. A separate answer sheet is provided for each section of the tests.

The Chicago Mathematics Program

The Chicago public schools calculator-supported mathematics assessment program was designed to accompany the institution of its new Comprehensive Mathematics Program (Chicago Public Schools, 1987). This program was designed to enable students to master those competencies and skills which reflect the learning goals and objectives developed by the Illinois State Board of Education.

Instruction to promote the mastery of computational skills, the development of problem-solving strategies, the use of the calculator, and the acquisition of the language of mathematics is emphasized. The emphasis on using calculators is in keeping with the Illinois State Board of Education's policy on calculators, which states on page 5 of *State Goals for Learning and Sample Learning Objectives: Mathematics* (n.d.):

Students should be able to perform some skills mentally, do some with paper and pencil, and do some using the appropriate technology.

The Chicago Public School's mathematics program for the elementary school is defined in three handbooks. These handbooks identify specific objectives that are to be achieved each quarter. At each grade level, objectives are accompanied by examples which clarify the meaning of the objective and models the level of proficiency that will be assessed. Criterion-referenced tests, which assess mastery of the objectives of the Comprehensive Mathematics Program, are administered each quarter to all students in grades 1 through 8.

Developing Items for Calculator-Supported Assessment

One of the greatest challenges faced in instituting calculator-supported assessment was the development of appropriate test items. Items that are traditionally included on criterion-referenced tests focus primarily on computational skills. The challenge was to develop items that would assess skills beyond the computational skills level. The following criteria were defined for item development: (i) The item must test achievement of the stated objective; (ii) The question must be posed in such a way that more than pushing buttons is required to get a correct answer; (iii) The arithmetic computations section of the test must go beyond simple mathematics symbol manipulations to assess mathematics principles involved in computational processes.

To be honest, we set out to make items in this section "calculator-proof." Thus, the item had to go beyond simple mathematics computations at the "button-pushing level."

Building understanding of calculator features

Objectives and accompanying test items such as Figures 7.1A and B (see page 115) address students' understanding of some of the special features of their calculators.

Assessing skill proficiency with and without calculator support

Students are expected to be proficient in mathematics computations with and without calculator support. The criterion-referenced tests include items to measure computational proficiencies at both levels. To demonstrate calculator proficiency, students must write calculator sequences that could be used with computations involved in problems. To demonstrate proficiency with mathematics computation, students must be able to perform mathematics computations without a calculator. The objectives, examples, and test items that model these two levels of computational expectations are shown on pages 115–117 (Figures 7.2A, B, C, D, E, F, and G).

Calculator support for problem solving

Calculators are available to use with computations involved in problem solving. Students are expected to make computational decisions, deciding whether or not to use calculators with objectives such as those in Figures 7.3A, B, C, and D (see page 118).

Assessing use of calculators with problems involving formulas

Students are expected to be able to use calculators to support problem solving involving formulas. Objectives, examples, and related test items focus on the students' ability to translate a word problem into a formula and to relate the formula to an appropriate calculator sequence. The objectives, examples, and test items on page 118 (Figures 7.4A–D) model the school district's levels of expectation.

Integrating Assessment and Instruction

The calculator allows very effectively integration of instruction and assessment. The disadvantage (or advantage) of such assessments is that it is difficult (or impossible) to distinguish instruction from assessment.

Consider the following problem:

Find the smallest whole number between 36 and 49 with a square root closer to 7 than to 6. Record your results of calculator trials on the chart below.

N	Square root of N
36	6
49	7

Problems such as this one could be used to assess students' levels of proficiency in the following areas: (i) understanding the meaning of square root, (ii) reading decimals, (iii) rounding decimals, (iv) comparing decimals, (v) the magnitude of decimals, (vi) decision making in mathematics, and (vii) reasoning.

Suppose a student's choices are those listed below in the order given.

N	Square root of N
36	6
37	6.0827
42	6.4807407
48	6.9282032
49	7

At this point, the assessor could ask the student why 48 was chosen after 42. This could lead to a discussion which could identify strengths and weaknesses in understanding the meaning of square root, reading decimals, rounding decimals, comparing decimals, the magnitude of decimals, decision making in mathematics, and reasoning.

Students could be given the following calculator sequence (some calculators have a 1/x key for INV):

$$\Box \quad Y^x \quad 3 \quad INV \quad =$$

They could be asked to substitute the numbers 1, 2, 3, 4, and 5 in the box and record their answers on a chart such as the one following.

Number	Display
1	1
2	1.2599211
3	1.4422496
4	1.5874011
5	1.709976

Directing students to identify several numbers for which this sequence will lead to a whole number facilitates assessing the students' understanding in several areas. I have found that many people who learned

Problem	Calculator sequence	Concept
1. 60 + 20% of 60 = N	60 + 20 %	Percent of increase
2. 60 - 20% of 60 = N	60 - 20 %	Percent of decrease
3. 60 x 20% = N	60 x 20 %	Meaning of percent
4. 60 ÷ 20% = N	60 ÷ 20 %	Percent and calculator order of operations

Figure 7.5. Percent problems, calculator sequences, and concepts.

cube roots mechanically (including some people with mathematics backgrounds) do not recognize the results as the cube root of the number given. I have also seen people who, after discovering that 8 satisfies the conditions of the calculator sequence and tested 64 and found out that it also satisfies the condition, assumed that the numbers would be the square of the last number found. As assessor could probe further by asking the student to substitute 27 into the equation and asking whether or not this fits their rule.

The problem at the top of this page demonstrates the value of the calculator in assessing student's understanding of percent applications. Consider the problems and the related calculator sequences shown.

By asking the question, "What is the calculator doing?" an assessor can determine a student's level of understanding of the meaning of percent (problem 3), percent of increase (problem 1), percent of decrease (problem 2), and percent and the order of operations on the scientific calculator (problem 4). By assessing these concepts as a unit instead of in isolation, one can determine whether students have mechanical mastery of percent concepts or have mastered them at a level at which they can apply them and relate them to each other.

The *Curriculum and Evaluation Standards for School Mathematics* states:

Assessment, in its turn, is conceived of as an integral element of instruction, a method of generating valuable information for both student and teacher. As instruction explores content through an array of problem-solving situations, so too must assessment.

References

Chicago Public Schools, Board of Education. (1987). *Implementation handbook for the comprehensive mathematics program, grades 7-8.* Chicago: Board of Education.

Illinois State Board of Education, Department of School Improvement Services. (n.d.) *State goals for learning and sample learning objectives: Mathematics, grades 3, 6, 8, 10, 12.* Springfield: State of Illinois.

National Council of Teachers of Mathematics. (1986). *A position statement on calculators in mathematics classroom.* Reston, VA: NCTM.

Figure 7.1A

Objective: Recognize how calculators round or truncate numbers.

Example: Read the word problem: Hamid divided 2.00 by 3 on his calculator by using the following key sequence:

Hamid's answer was

0.6666667

Rupa also divided 2.00 by 3 on her calculator by using the same key sequence:

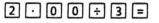

Rupa's answer was

0.6666666

What is the probable reason for the difference between Hamid's and Rupa's answers?

Answer: Hamid's calculator rounds off the last digit that is visible at the right of the display. Rupa's calculator does not round off the answers. Her calculator truncates (eliminates) all digits to the right of the last digit that is visible on the display.

Figure 7.1B

Test item: Choose the answer that should appear on the display of a calculator that *rounds* numbers.

$8 \div 9 =$
 a) 0.8888888
 b) 0.8888889
 c) 0.8888899
 d) 0.8888999

Test item: Choose the answer that should appear on the display of a calculator that *truncates* numbers.

$5 \div 3 =$
 a) 1.6666666
 b) 1.6666667
 c) 1.67
 d) 1.7

Figure 7.2A

Objective: Demonstrate the use of a calculator to solve word problems involving the addition and subtraction of integers.

Example: Read the word problem:
A department store experienced an eight-cent loss on every beach ball sold but had a four-cent profit on every baseball sold. If 350 beach balls and 475 baseballs were sold during a two-month period, what was the store's profit or loss on these two items?

The answer is a loss of $9.00.

Demonstrate a set of key sequences that could be used on a calculator to arrive at the answer.

Sample answer:

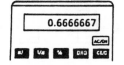

Figure 7.2B

Test item: The temperature at 9 a.m. was -12°C. By 1 p.m., the temperature had risen 10 degrees. By 5 p.m., the temperature had risen another 3 degrees. However, by 9 p.m., the temperature had fallen 5 degrees. What was the temperature at 9 p.m.?

How could you use your calculator to find the answer?

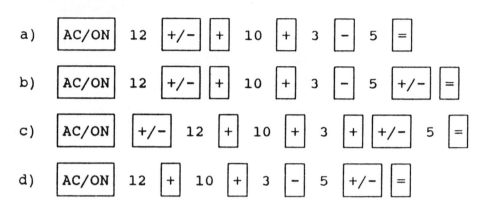

Figure 7.2C

Objective: Find small integer powers (-9 to 9) of positive numbers.

Example: Find the values without using a calculator.

a) 2^3
b) 2^{-3}

Answer:

a) $2 \times 2 \times 2 = 8$

b) $\dfrac{1}{2 \times 2 \times 2}$ = $\dfrac{1}{8}$ = 0.125

Find the values by using a calculator.

a) 6.13^7
b) 6.13^{-7}

Demonstrate key sequences that could be used on a calculator to arrive at the answers.

Sample answer:

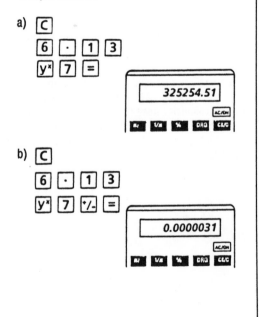

Figure 7.2D

Objective: Demonstrate the use of the calculator to solve word problems involving small integer powers (-9 to 9) for positive numbers.

Example: Read the word problem: Lucia has been collecting stamps from around the world since she was 9 years old. She collected 5 stamps the first year and was able to triple the size of her collection each succeeding year. If her collection continues to grow at that same rate, how many stamps will she have when she is 14 years of age?

Lucia made a chart to find her answer.

Age	9	10	11	12	13	14
Stamps	5	15	45	135	405	1215

The answer is 1215 stamps.

Demonstrate a key sequence that could be used on a calculator to arrive at the answer.

Sample answer:

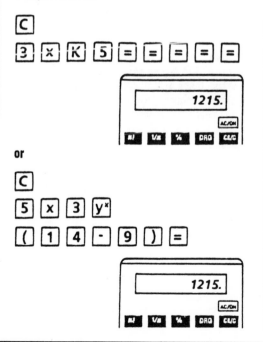

or

Figure 7.2E

Test item: Allan has 2 dollars. Bonnie has twice as much money as Alan. Carlos has twice as much money as Bonnie. Danny has twice as much money as Carlos. Evonne has twice as much money as Danny. Felicia has twice as much money as Evonne. How much money does Felicia have?

How could you use your calculator to find the answer?

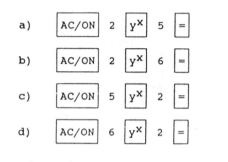

Figure 7.2F

Objective: Add and subtract integers.

Example: Determine the missing integer in each equation.

    a) $^-80 + {^+}108 = n$

    b) $^-43 - {^-}40 = n$

Answer: a) $^+28$ b) $^-3$

Figure 7.2G

Test item: Solve the equation for n.

    $n = {^+}86 - {^-}44$

Answer:

    a) $^-42$ b) 42 c) $^-130$ d) 130

Figure 7.3A

Objective: Solve problems involving tips.

Example: Read the word problem:
Show the two-step process that is needed to solve the problem. Then find the answer.

The restaurant bill for Jill's birthday dinner totaled $25.80. The usual tip for the waiter is 15% of the total bill. What was the total expense of the dinner (bill + tip)?

Answer:

Step 1: $25.80 x .15 = n; n = $3.87

Step 2: $25.80 + $3.87 = t; t = $29.67

The total bill was $29.67.

Figure 7.3B

Test item: A family's bill for dinner was $24.60. How much is a 15% tip?

a) $0.15

b) $0.36

c) $1.50

d) $3.69

Figure 7.3C

Objective: Solve problems involving a commission.

Example: Read the word problem and the information in the chart to solve the problem:
A sales person earns 12.5% commission on all sales. During the super sale, a salesperson sold 1 coat, 2 sweaters, and 2 skirts. How much commission did the salesperson earn?

SUPER SALE!!	
Wool coat	$79.50
All sweaters	$8.98
Selected skirts	$11.77

Answer:
[$79.50 + 2($8.98) + 2($11.77)] x 0.125

= $15.125, or

= $15.13 (rounded to the nearest cent).

Figure 7.3D

Test item: Ms. Sherman sells carpeting. Her salary is $700 a month plus 3% commission on her sales. If she sold $14,870 worth of carpeting in January, how much money did Ms. Sherman earn that month?

a) $446.10

b) $721.00

c) $1,146.10

d) $15,570.00

Figure 7.4A

Objective: Demonstrate the use of a calculator to solve word problems that involve formulas containing cubic terms.

Example: Read and solve the word problem: A spherical balloon is inflated so that its radius is 20 cm. Use the formula to find the volume of the balloon. Round the answer to the nearest hundredth.

$$V = \frac{4}{3} \pi r^3$$

The answer is 33,510.32 cubic centimeters rounded to the nearest hundredth.

Demonstrate a key sequence that could be used on a calculator to arrive at the answer.

Sample answer:

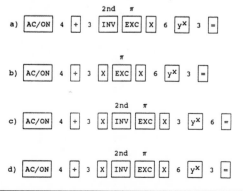

Figure 7.4B

Test Item: The radius of a sphere is 6 inches. Use the formula $V = 4/3 \, \pi r^3$ to find the volume of the sphere.

How could you use your calculator to find the answer?

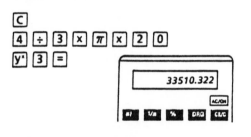

a) AC/ON 4 + 3 INV EXC X 6 yˣ 3 =   (2nd π)

b) AC/ON 4 ÷ 3 X EXC X 6 yˣ 3 =   (π)

c) AC/ON 4 + 3 X INV EXC X 3 yˣ 6 =   (2nd π)

d) AC/ON 4 + 3 X INV EXC X 6 yˣ 3 =   (2nd π)

Figure 7.4C

Objective: Demonstrate the use of the calculator to solve word problems involving formulas containing square roots.

Example: Read the word problem: The shape of the Brown's living room floor is a square. The area of the room is 73 square meters. Use the formula $s = \sqrt{A}$ to find the length of each side of the room.

The answer is 8.5440037 meters.

Demonstrate a key sequence that could be used on a calculator to arrive at the answer.

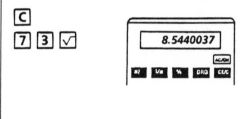

Figure 7.4D

Test item: The top of a table has the shape of a square and an area of 9,604 cm². Use the formula $s = \sqrt{A}$ to find the length of each side of the table top.

How could you use your scientific calculator to find the answer?

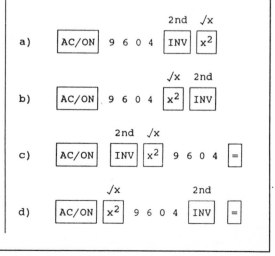

a) AC/ON 9 6 0 4 INV x²   (2nd √x)

b) AC/ON . 9 6 0 4 x² INV   (√x 2nd)

c) AC/ON INV x² 9 6 0 4 =   (2nd √x)

d) AC/ON x² 9 6 0 4 INV =   (√x 2nd)

8

Advances in Computer-Based Mathematics Assessment

JOSEPH I. LIPSON, JOSEPH FALETTI, and MICHAEL E. MARTINEZ

Mathematician/philosopher Alfred North Whitehead has been credited with saying that civilization advances by extending the number of important operations which we can perform without thinking of them. In this century, an insufficiently appreciated task of education has become to extend automated skill in key operations to all citizens. As a consequence, it becomes the task of educational assessment to determine the presence of these automated skills. Often, students will demonstrate these automated operations in familiar educational settings with traditional assessments, but are unable to apply them to noneducational situations (for example, see Erlwanger, 1973; Rosnick and Clement, 1980; diSessa, 1982; McCloskey et al., 1980; and a general discussion in Davis, 1984).

Many educational reformers believe that only through radical restructuring can our schools show the needed improvement. However, two narrow external pressures can be applied to improve education significantly without the complete restructuring requested by some reformers (Holden, 1989).

The first is curriculum change. By changing the courses offered and deciding which of these courses are required, we change the concepts and skills that students should learn. Regardless of course quality, students required to take several years of intensive mathematics will be different from students for whom advanced mathematics courses are either elective or not available. A significant tool for such change is the creation of new standards such as those of the National Council of Teachers of Mathematics (NCTM, 1989).

A second prod for reform in education is a change in tests (Fredericksen, 1984). Tests can influence course goals; a new consensus on tests often crystallizes an educational problem in the public mind and leads to curriculum change. Recent examples include the National Geographic Society's test of geography knowledge, Hirsch's (1987) specification of cultural literacy (although not strictly a test), and the National Assessment of Educational Progress (NAEP), all of which have fueled curriculum debate. In the rhetoric of education, the evidence from tests is often persuasive.

In this chapter, we explore ways of improving mathematics assessment so that it better supports the goals of education and the needs of students. Many of these improvements require operations that can only be done efficiently with a computer, so the design and architecture of a computer-based system for improved assessment is also discussed. In view of the increasing availability of low-cost, powerful computer workstations, and developments in cognitive science research, we will explore the possibility of large-scale mathematics assessment/instruction involving "intelligent" computer-based systems. Given these tools, we can, at the same time, make a major change in our conception of what assessment in all domains can and ought to be.

The Inadequacy of Current Mathematics Assessment

There are many inadequacies in current mathematics assessment methods that we seek to remedy. We divide them into three groups: negative effects on teaching and learning, inadequacies in the kinds of cognitive activities they require of students, and inadequacies in the kinds of information gathered and reported.

Effects of tests in teaching and learning

Norman Fredericksen (1984) has observed that

Multiple choice tests tend not to measure the more complex cognitive abilities.... It is suggested that the greater cost of tests in other formats might be justified by their value for instruction—to encourage the teaching of higher level cognitive skills and to provide practice with feedback.

The implication of this observation is that the design of a testing system should take into account the effects and uses of the test by a variety of audiences. When we do not consider the full range of interactions, we risk doing more harm than good. For example, if students spend a significant amount of time learning how to take tests rather than learning the material being tested, the net result might be improved scores and reduced knowledge and skill.

Constriction of curricula ("teaching to the test"). The most unfortunate recent product of tests has resulted from demands of greater accountability for schools and teachers. As tests are used more frequently as measures of teacher performance—rather than of individual students' performance—teachers are quite naturally motivated to teach the material they know will be on the tests in the manner that it will be tested, sometimes called teaching to the test. This results in tests becoming the implicit definition of the curriculum. However, tests do not cover the entire spectrum of

material or modes of learning or performance in the areas they test. Thus, a side-effect of teaching to the test is that the curriculum shrinks to fit the test. Since tests' results are only intended to correlate with learning resulting from a healthy instructional environment, they do not pick up this radical change in instructional environment.

As curricula constrict to fit tests, a second related effect also occurs. Since tests are often designed to be independent of curricula, they represent the intersection of many curricula. As a result, the richness that has characterized curricula in the past is often being eliminated. Thus, the de facto curricula are shrinking and merging—toward their intersection as defined by commonly used tests.

To counter this effect, tests must be expanded to encompass the wide variety of activities and contexts that curricula should include and the wide variety of content that curricula might include. This would allow tests to provide some indication of the border between what a student knows and the vast universe of knowledge that has not yet been learned by a student or class. It would also allow special or unusual content that a particular teacher or curriculum includes to be assessed.

At the same time that we are working on redesigning testing methods, of course, there are valuable efforts being made to improve the curriculum directly, both by creating new standards and new textbook series. The most prominent of the new standards in mathematics are those of the National Council of Teachers of Mathematics (NCTM, 1989). While much of the work reported here occurred in parallel with this effort, we are quite pleased with the considerable degree of overlap between their evaluation standards and the changes that we are proposing. In particular, the NCTM agrees that "Tests also must change because they are one way of communicating what is important for students to know. The tested curriculum can strongly influence what students are taught."

Impoverishment of the testing situation. A related inadequacy of current tests is the growing gap between the testing context and the learning context. This has always been the case to a certain extent in science classrooms, where tests have traditionally contrasted sharply with the laboratories that incorporate active involvement with batteries and wires or test tubes and chemical substances. As mathematics classrooms are enhanced with new tools, such as calculators and computers, and with new learning modes such as cooperative learning, this gap is becoming more appreciated by mathematics teachers and educators. The implications are worse in mathematics, however, because calculators and computers are more inseparably integrated into the mathematics classroom's daily activities than science laboratories have traditionally been. In this context, the traditional "clear off your desk and take out your pencil" becomes akin to asking a carpenter to "put away your power saw and nail gun and take out a hand saw and hammer." As some of the more advanced computer-based tools to be described later are integrated into classrooms, pencils will be gradually reduced from the already impoverished status of hand saw to that of a pocket knife.

In order to keep up with advances in instructional tools, these tools must be incorporated into tests and made an integral part of them. Thus, tests of material that benefits from the use of such tools must be administered on the same computers and intimately integrated with those tools.

Direct and anticipation effects. There are many ways in which the kinds of tests that are given affect how students and teachers approach the material to be tested. We will only mention a few.

Among the intended or implicit effects of tests are (i) to motivate students to study; (ii) to inform students of how much they know, and whether, in the view of the adult world, their study strategies have been effective; and (iii) to define an achievement hierarchy of students. Students may interpret the results of tests in ways that have important educational implications, leading them to study more intensively or to decide they have no aptitude for mathematics and therefore avoid it in the future. While we cannot predict all the ways that tests can be used or misused (Hoffman, 1962), we can make the uses of tests part of the design criteria for the system.

Another class of effects are anticipation effects (Lundeberge and Fox, 1989). Anticipation of a test by students and teachers influen-

ces the way that students and teachers spend the resources at their disposal. Student resources include mental effort, time, or money for tutors or supplementary study materials. Teacher resources include class time and the fund of specific activities available. While preparing students for a high-stakes, externally administered test — especially when they are also used to evaluate teachers and schools — teachers' self-interest increases the likelihood that they will modify their classroom activities and assignments in anticipation of a test.

One of the authors (Lipson, 1988) has argued that in addition to conscious decisions made in anticipation of a specific test and test format, there may be significant subconscious effects. The thesis is that once an important goal is clearly held in mind, this alters the way that information is organized and held in mind. As Dr. Pandey of the California Assessment Program asked at a testing conference, "Does multiple-choice testing lead to multiple-choice thinking?"

There are many other ways in which students, teachers, parents, administrators, and society may react to the results of tests. In spite of the manifold effects of tests, we have not seen a comprehensive attempt to specify and develop the kind of testing system necessary to optimize the positive effects and to minimize the negative effects of tests. Tests ideally should be designed so that these reactions will induce choices beneficial to learning.

Cognitive inadequacies

As cognitive science advances, much is being learned about the processes underlying mathematics learning. A summary of the work in this field would be impossible here, although a good starting point is provided by Davis (1984). Suffice it to say that existing curricula and tests are inadequate in light of this research, both in implicit student models and in the cognitive processes they engender.

Choice responses versus constructed responses. One possible near-term improvement in mathematics assessment lies in test items that require construction of answers. Constructed responses, in principle, are easier to score in mathematics than in other subject areas. The Educational Testing Service (ETS) is currently experimenting with ways in which students can indicate constructed numeric responses on paper-and-pencil tests. With computer-delivered instruction, the difficulties are much reduced, of course. However, formulaic responses may be difficult for students to generate on a computer.

Arguments can be made that constructed responses tap different sorts of skills than do responses chosen from a list. In tasks of figural reasoning, Snow (1987), for example, found that high-ability examinees construct answers mentally before they look at the options. In contrast, lower-ability students tend to work backwards from the options provided to arrive at their choice. To our knowledge, little work has been done in this area for mathematics, but it is plausible that calling for constructed responses is more likely to elicit productive thinking, rather than encouraging students to use a strategy better characterized by test-wiseness than mathematics problem-solving strategy. Snow and Peterson (1985) report several other effects of constructed responses, including a result from Schmitt and Crocker (1981) which found that students with high test anxiety scored significantly worse on a constructed response test, while students with low test anxiety scored significantly better. Such results make it clear that more research is needed on the underlying cognitive skills of new item types. Perhaps most significantly, constructed items are well suited to diagnosing students' error patterns than are items that require selection of a correct response from among options (Birenbaum & Tatsuoka, 1987).

"Success" without understanding. The importance of diagnosis of error patterns is demonstrated in so-called "disaster studies." In these studies, students judged successful in a domain in educational settings with traditional assessments fail miserably in non-academic applications of basic principles, in many cases, giving the same answers that untrained novices would give (see Erlwanger, 1973; Rosnick and Clement, 1980; diSessa, 1982; McCloskey et al., 1980; and a general discussion in Davis, 1984). Upon further exploration, it was found that the students were capable of applying the limited algorithms they were taught to the limited contexts they were taught about, but had no deep under-

standing of the content underlying these algorithms. Perhaps even more disturbing are instances in which both students and experts take such a narrow mathematical view of a problem that they infer the intended method to be used and use it without noticing the common-sense complexities of a problem.

One such example arose on a recent SAT test in which the following "rolling circle question" appeared:

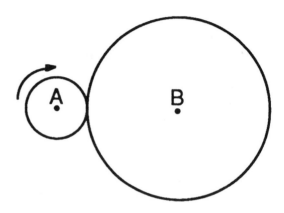

In the figure above, the radius of circle A is ⅓ the radius of circle B. Starting from the position shown in the figure, circle A rolls along the circumference of circle B. After how many revolutions will the center of circle A first reach its starting point?
(A) 3⁄2   (B) 3   (C) 6   (D) 9⁄2   (E) 9

The answer which the item designers indicated was the "key" was (B), but interestingly, this item was actually missing the "correct" answer, which is 4. Wainer et al. (1983) report that the choice of the key, despite its incorrectness, correlated highly with the overall test score. Most attempts by Wainer and his colleagues to elicit the correct response from various experts and students at the appropriate level of learning failed. The item was assumed by the designers, experts, and students to be an application of the principle that the ratio of the circumference of two circles equals the ratio of their respective radii. Yet, a full understanding (apparently rare) of what is happening must factor in the motion around the larger circle. Even the "best" students (as measured by this test and Wainer's studies) were perfectly happy with the key.

A variation on "success without understanding" might be called "failure due to lack of understanding." The results of the DEBUGGY work on multicolumn subtraction (Brown and Burton, 1978; Brown and VanLehn, 1980; and VanLehn, 1982) are an example. Analysis of many of the bugs, especially those involving borrowing, makes it clear that students are attempting to reproduce a memorized algorithm without understanding of the source of that algorithm in the notion of place-value that underlies our number system. This raises the question, unanswered by standard subtraction tests, of whether students performing the algorithm correctly understand any better why it works.

Discrete exercises and context-free items. The episodic nature of tests constructed from discrete, disconnected, and often context-free exercises is also cognitively inadequate. Mathematics arises frequently in everyday life, but it seldom presents itself in the compact form that we find in typical school mathematics problems. Rather, the normal experience is to have extraneous information, several related and interdependent problems to solve, and a range within "reasonable" answers must fall. Tests should present similar situations. In such a context, students will be required to determine what information is relevant. They will be less inclined to apply memorized algorithms by rote and more inclined to use common-sense knowledge to judge the reasonableness of their answers.

The DEBUGGY work again provides an example. Many of the bugs result in answers that clearly would be wrong if students considered the meanings of subtraction and of the numbers involved. But such tests take students out of the "understanding" mode and instead put them into a "subtraction-machine" mode.

Choice and interpretation of distractors. An important inadequacy in current assessment is the kinds of information gleaned from individual responses. Typically, a response is scored either right or wrong. But there can be many causes for wrong answers, and each wrong answer can tell us something about the student's knowledge. The rolling circle question (Wainer et al., 1983) is a perfect example of this. If the correct answer had been included, then there is no evidence that any student would have computed the correct answer, and the obvious inference would be that students do not understand the underlying principle (or that the problem was bad, as will be discussed). An alternative is to glean more information from the answer. If the student gets the previous "key" of 3, then we can infer that he/she understands about the ratios of circumferences. If the student gets the correct answer of 4, then *in addition,* we can infer that he/she completely understands the problem and not just the individual mathematics fact underlying it.

Given the fact that few if any students would choose the correct answer, the rolling circle question with the correct answer as one of the choices would have been eliminated during pretesting since it would not correlate well with overall performance on the test. Similarly, when there are common misconceptions held by otherwise successful students, an item that taps that misconception would typically eliminate a problem. The net result is that errors that are very common among generally successful students are never caught. Thus, students who, in fact, have room for improvement receive scores that say they were (near) perfect, thus sending the wrong message.

Thus, in general, we can gain insight from individual responses into the details of a student's knowledge and understanding. However, such inferences must be followed up and verified by other problems, such as presenting psychometric problems, as will be discussed. In a paper-and-pencil test with single-score reporting, this information would be too voluminous to use. But with computer-based tests, there is the potential to develop and use this information to inform students' and teachers' next steps in instruction.

How Technology Can Help

From our perspective, the future of computer-based mathematics assessment lies (or should lie) in a transition from multiple-choice formats with a single correct choice to open-ended or constructed-response questions and problems that test higher-level knowledge and skill (NCTM, 1989).

In the long term, we wish to glean as much

information as we can from students' performances and use that information to support teachers' and students' decisions on what to teach next and how to teach it. Thus, we want to know many things about a student's knowledge and skill:

(i) What comes to his/her mind (associations) when a given question is asked, when a word is heard, when a problem is posed, when a picture is seen? These questions can reveal the meaning of mathematical concepts for the student.

(ii) How does a student represent a problem? What external aids (e.g., figures) does the student create? How does the student organize his/her work in order to be able to retrace steps when a problem arises?

(iii) How does the context and/or area of application of a problem affect the student's response?

(iv) How does a student spontaneously organize knowledge that is presented?

(v) What algorthmic and heuristic procedures are automated so that their use does not interfere with concentration on the meaning of the problem?

(vi) What reasoning processes does the student use when faced with a nontrivial problem? How does the student respond to problems whose statements have missing information, or excess information?

(vii) How does a student estimate an answer in order to check on the reasonableness of a result? How does a student check his/her work? Does the student try to generate alternate forms of a solution in order to explore the "solution space" and to compare the positive and negative aspects of a given type of solution?

(viii) How does the student respond to difficulties? How does the student deploy his/her mental resources when a roadblock to a solution is encountered?

(ix) What mathematical knowledge does the student have that is outside the curriculum?

(x) How does the student organize his/her effort for problems that require extended effort?

(xi) How is a student's general world knowledge used while solving mathematics problems or during mathematical thought?

(xii) When asked, how does the student construct questions to explore the knowledge of another student?

The implication of this list is that we need a testing system that can present many different kinds of problems to a student, and that can track and interpret not only the answer, but the intermediate steps in a solution attempt. The system also needs to (i) ask questions that probe the reasoning process; (ii) ask follow-up questions; (iii) present graphic material as part of the problem statement; and (iv) permit the student to respond with drawings, as well as words and symbols.

These attributes have important implications for the design of the computer-based system we have in mind. However, as we shall see, we also need to (i) build a model of the student's ability, strategies, intentions, and preferences, (ii) interpret the student's responses, (iii) simulate mathematical models and applications, (iv) have a strategy (theory) of the appropriate next question, and (v) construct a variety of useful reports for different audiences.

We envision a system with the following capabilities and characteristics: (i) generation and presentation of a wide range of item types; (ii) choice of items based on (a) the immediate past performance of the student, (b) the cumulative history of the student's work and preferences, and (c) the conceptual structure of mathematics; (iii) analysis and interpretation of the student's work for more than merely the correctness or incorrectness of the answer; (iv) maintenance of a dynamic model and representation of the student's knowledge and skill; (v) incorporation of a knowledge base of mathematics relevant to the K–12 curriculum and applications outside of the school context. Such a knowledge base could be used to analyze the relationships among mathematical concepts as a basis for choice of test items; (vi) provision of feedback in the form of a progress map (Bunderson, Inouye, and Olsen, 1989; Bunderson and Forehand, 1989) depicting the student's learning over time, suggesting possible areas for additional study, and mapping out the "landscape" ahead of the knowledge to be gained, including areas of probable difficulty; (vii) provision of timely, digestible, and useful reports to teachers (and parents) on current work of individual students; (viii) provision of reports that aggregate information about subgroups of students for appropriate audiences; and (ix) suggestions for instruction on the basis of the aggregated information about the performance of a class.

While implementation of these pose a great challenge, we have thus far made progress on the first five items.

Recent progress in computer-based assessment

The advent of computers has generated a multitude of educational software products having instruction or learning as their focus but with assessment components. We will explore these briefly to set a context for the wide variety of methods that could be added to a computer-based assessment system. For a discussion of previous work on computer-based assessment with assessment as its primary focus, see Bunderson, Inouye, and Olsen (1989).

Computer-aided instruction as a testing system. At present, computer-assisted testing is available only in limited forms, although test items and test-like activities are usually a major part of computer-aided instruction (CAI). In fact, much CAI in mathematics uses a drill-and-practice format in which a sequence of problems is posed. If a student gets the item correct, he/she is presented with another problem. In this way, the student moves through a collection of lessons according to some sequential strategy.

CAI as a testing system provides limited information regarding a student's knowledge, abilities, and strategies. Since the goal is learning, the performance is an uncertain mixture of the student's knowledge and of help provided by the CAI lesson. As such, it is unlikely to reveal what the student knows from experience and interests outside of the classroom.

Task-oriented instructional feedback. In contrast to CAI's focus on instruction, other educational computer programs present a series of problems of increasing difficulty, followed by feedback on the quality of their solution. Typically, little explicit instruction is given. Such programs usually support learning in a narrow domain. "Rocky's Boots," for example, requires the user to learn and apply logical rules to sort objects on the computer screen. "Algebra Arcade" and related programs require the student to construct functions passing through a given set of points.

While few of these software packages have a solid research base, they seem to provide challenge and feedback lacking in typical instruction and assessment. The tasks are usually more complex than typical test items and the contexts are often more motivating. The feedback for both success and error is usually immediate. Because of the nature of the score and the increasing complexity of problems, the students are aware of their progress. Often, the potential for progress seems open-ended, reinforcing the important notion that there is always much more to learn.

Computer adaptive testing. The idea behind computer adaptive testing (CAT) is relatively simple (Wise and Plake, 1989). Test items in a domain are ordered according to difficulty based on prior testing. If the system has no prior rating of the student, the first item the student encounters is of medium difficulty. If the student gets the item correct, the next item is more difficult. Gradually, the system homes in on a rating for the student. As the use of an item increases for a particular population, its difficulty rating is continually updated. Typically, CAT arrives at a statistically reliable score for a student in about one-half to one-third the time of a conventional test in which the student is asked to answer all items.

One limitation of CAT is that items are usually ordered only according to difficulty. Psychometric research is developing methods that will allow items to be grouped according to attributes other than difficulty, such as complexity, memory load need to change representations, and content. If this line of research reaches fruition, CAT could be much more useful for the goals we now seek of mathematics instruction.

The CAT method is most useful when items are closely related to each other in the content and skills required and when a consistent difficulty ordering can be made. Such tests for a variety of subdomains of knowledge and skill can be combined. A related approach uses tests composed of "testlets" (Wainer and Kiely, 1987). A testlet contains a small number of related items, often drawing on a single problem context, that test a specific area of knowledge and skill, such as solving rate and distance problems. Psychometric analysis allows us to determine the minimum number of items on a testlet that reliably test the area of interest.

Other potential contributions of technology

Richer representations via computer. In evaluating the worth of computer-delivered assessment, a point that should not be overlooked is that the computer can offer a much richer context for problems than can printed tests. When using print, it is expensive to present drawings, graphs, and figures as part of the problem statement. As a result, there is a tendency to use words and compact symbol systems. Computer systems make it easier and less expensive to use rich problem representations. Presentations can exploit the computer's capability to use graphics, color, animation, and even photographically realistic images. Problems that have a common conceptual basis can be modified to contexts that are familiar or of interest to the student. For example, problems in proportion can be expressed in the context of the growth of pets or the relative costs of buying cassettes versus compact discs.

The preceding is not meant to suggest that we should always cater to areas of student interest or familiarity. Familiarity, interest, and context appear to change the information-processing load and strategy of the student. One variable to be examined in assessment is the extent to which different contexts evoke different responses. Eventually, we would like our students to have sufficient knowledge, strategies, knowledge of procedures, and power of reasoning to be able to provide their own contexts for a problem, to see the similarities of a problem that is posed in different contexts or in a context-free form.

Richer feedback via computer. When a student takes a printed test, the typical feedback is a much-delayed score and a ranking. In contrast, the computer has the potential to dramatically represent the match between a problem solution and the statement of the problem. For example, if we ask the student to find how much paint is needed to cover a wall, the computer can graphically show the difference between the given area and the area covered by the paint stated as the solution. Or when a student gives a poorly proportioned response to a well-proportioned question about a person's running speed, the unreasonableness of his/her response could be made explicit by comparing it to the current world's record.

Constructed or figural responses

One limitation of current test items in mathematics is that while line drawings, diagrams, and graphs can be presented as part of a problem, figures drawn or modified by the student cannot be scored automatically. Two of the authors (Martinez and Faletti) are currently developing a capability for reading and interpreting figural responses. This project has the goal of adapting techniques used for scoring other kinds of tests. Some of the new item types and the cognitive analysis of students' methods for generating responses to them is being transferred to the development of computer-based items.

Multiple-step problem solving/ "cognitive cryptography"

To move away from discrete and context-free items, problems can be designed to provide extraneous information and require multiple steps. This presents a significant challenge for the scoring algorithms. The goal is to take student performances and interpret the underlying cognitive models that generate them, thus performing a kind of "cognitive cryptography," decoding the knowledge underlying the performance. In this section, we examine some of the ways in which the students' knowledge can be explored to find an underlying structure.

Two-step items. One of the limitations of multiple choice is that we have no information about the reason for a student's choice of alternatives. With follow-up items based on the particular choice, we could obtain information about students' reasoning, while possibly decreasing the guessing. Inappropriate reasons would be reflected in a modified score and could be another resource for diagnosis of student's strengths and weaknesses.

Adding follow-up items to printed tests is impractical, since they would multiply the size of a test. However, if the items and follow-up items are presented by computer, this difficulty is avoided. In addition, follow-up items could be invaluable as instructional aids. Davis (1984) reports on little-known work by

David Page on "standard wrong answers" among students learning arithmetic. For example, students asked to multiply will respond with the sum, students asked to take an exponent will multiply, and students asked to divide by 1/2 will multiply by 1/2. Page found that when a student had done this, asking the problem the student *answered* (instead of the one asked) as a follow-up question would cause the student to notice and repair his/her earlier error. Davis reports that the following actual dialogue is quite typical:

TEACHER: How much is seven times seven?

STUDENT (in grade seven): Fourteen.

TEACHER: How much is seven plus seven?

STUDENT: Oh! It should be forty-nine!

This is the kind of follow-up that computers could quite easily generate, resulting in beneficial effects on the student's learning.

Mental models. As students learn, they acquire an increasingly complex network of knowledge, associations, images, skills, and expectations. One goal for computer-based assessment is to build a representation of the student's mental model in mathematics. Once assessment is defined in this way, we have a very different image of testing from the conventional one.

Instead of asking whether students can get the right answers, we must construct different questions that we wish to find the answers to. When confronted by a problem, how do students interpret and organize the information available? What strategies and tactics come to mind? What other information do they bring to bear on the problem? When students are told that a problem has an error in the solution, how do they go about finding that error?

These kinds of questions lead to a greater emphasis on qualitative and conceptual mathematics as the precursor and framework to the invocation of automated skills. It suggests that when a student displays a routine skill, we need to ask what may be called the "How-do-you-know?" question. For example, if a student has memorized that $7 \times 8 = 56$, it is appropriate to ask, "How do you know that 7

$\times 8 = 56$? How can you be sure that it isn't 57 or 58?" A satisfactory answer might be to say or demonstrate that one can make 7 rows of 8 objects that can then, in fact, be counted to total 56.

For most students, most of the time, an advanced testing system should require students to take real-world problems and then interpret and model these events and situations by means of mathematical concepts. Eventually the student attracted to mathematics may want to reason with mathematical concepts purely in abstraction, but most will benefit from a more problem-solving–oriented approach.

The interplay between ideas and their practical applications has great potential for greater understanding. In addition to the contextualization and motivation provided by real-world problems, they also allow more realistic complexities to be added, such as multistep solutions or information that is extraneous to the problem but appropriate to the situation. This also provides a context in which science and mathematics can be fruitfully integrated.

The use of mathematics in design problems and other, more complex multistep problems is also a potentially rich source of items. Such problems can be challenging, can draw upon the interests of the student, and can illustrate the value of sophisticated mathematical understanding. In addition, over the years, various projects have developed a fairly impressive collection of such problems.

Item generation. Given cognitive models of the knowledge underlying particular kinds of problems, computers could automatically generate items along with the inferences that can be made from the responses that those models would generate. For example, the DEBUGGY program (Brown and Burton, 1978) was capable of generating diagnostic tests for particular bugs in subtraction algorithms.

Challenges to Improving Computer-Based Mathematics Assessment

The kind of computer-based assessment described earlier is still confronted with many challenges and unsolved problems. We will discuss a few of the major ones briefly.

Cognitive modeling

Large gaps remain in our understanding of the cognitive models underlying students' learning and performance in mathematics. Thus, early computer-based assessment must focus on areas that mathematics education research has explored in detail and has begun to understand. Merely pulling together the existing research and incorporating it into a coherent model is a challenge. However, by taking a single (but broad enough) area of mathematics and developing a knowledge-based system with tools for representing these cognitive models and incorporating new facts into them, we will have a tool that can support its own extension as further research is done.

Psychometric modeling

As yet, no psychometric models exist for testing of the sophistication and diversity we envision. However, theoretical work has started at ETS to enable us to analyze a student's work and be able to state the degree of confidence we can have that a domain of knowledge has been mastered. First steps have already been made in combining various psychometric approaches with cognitive models, with progress reported by Yamamoto (1987), Gitomer and Yamamoto (1989), Masters and Mislevy (1988), and Tatsuoka (1983).

Design of a Computer-Based Assessment System

In this section, we present the design of a computer-based assessment system that is under development.

Shortened sights

Since it will be necessary to have a usable and testable system long before the full-blown system is complete, each component will begin as a limited and non-intelligent version that can gradually be expanded to encompass wider and more powerful functions as the necessary knowledge bases are developed. In particular, the range of inputs, events, or actions that each component will be able to handle will gradually expand through a continuum from simple to more complex.

For example, the kinds of items initially available will be ones whose answers can be constructed by the student and understood easily by the computer. The possible responses must be able to be matched to the underlying cognitive models and individual pieces of knowledge that students must have to generate them. As methods for eliciting additional kinds of inputs are developed, items using those inputs follow.

Knowledge bases

In its fully developed form, the system will require several knowledge bases, including the expert knowledge base, the student model, and the task knowledge base.

The expert knowledge base contains enough expert knowledge to generate correct responses to items and to construct a map of the knowledge and skill to be mastered. Ideally, this knowledge base will also contain enough information to drive the simulation of a particular domain.

The student model contains information gathered about a particular student. The nature and extent of its overlap with the mathematics expert knowledge base will inform the assessment component. Where there is no overlap, the system has not yet determined that the student knows this area. If the area has been explored before, then lack of overlap implies lack of knowledge by the student. Thus, the student model must compile a running knowledge base for each student of their (i) knowledge—facts that are correct (perhaps only for their level) that we have seen evidence that they know; (ii) lack of knowledge—facts that we have seen evidence that they do not know; and (iii) incorrect notions—false "facts" that we have seen evidence that they believe (including, but not limited to, well-known preconceptions or misconceptions).

The task knowledge base contains a large collection of mathematics problems. Each task will have associated responses (or classes of responses) that known student models might generate. Associated with each response will be a set of inferences about the student's probable model, which consist of knowledge, lack of knowledge, or incorrect notions. Tasks will be indexed by these inferences so that tasks capable of testing a particular piece of knowledge can be retrieved easily. In the early stages, items will be fairly static but, eventually, large numbers of specific instances of items should be produced from broader design criteria by item generators.

Modules

Given these three knowledge bases, the system will require the following components.

Task selection module. The goal of the task selection module is a sequence of tasks that is automatically constructed based upon the following subgoals: (i) support inferences of knowledge, (ii) support inferences of lack of knowledge, (iii) support inferences of incorrect notions, (iv) expand zone of the expert knowledge base that is covered by the student model in a reasonable fashion, (v) stop trying to support well-substantiated inferences, (vi) seek verification of inferences of knowledge after a delay, and (vii) seek verification of inferences of knowledge in new contexts.

The current version focuses on achieving the first four subgoals. The latter two subgoals will require new psychometric models for these new assessments and need further research.

Presentation module. The purpose of the presentation module is to control and coordinate output presented to the student. It will present pictures and word lists to be selected from and will gradually be expanded to provide tables and graphs, animations, learning progress maps, statistical and psychometric data, sound, and semantically generated natural language and speech.

Task interpretation module. The task interpretation module will gather the student's response to a task and use the knowledge associated with each response for the current task to update the student model.

Report module. The report module will match the current student model against the expert model to produce a summary of what the student knows, does not know, and has not yet demonstrated. The resulting report will be a learning progress map of those concepts and skills that the student has mastered (Bunder-

son, Inouye, and Olsen, 1989). The exact nature of the learning progress map remains undetermined and will be influenced strongly by the structure of the knowledge bases developed in the current phase.

Initial focus

Given the complexity of this design, current work is focusing on the construction of the three knowledge bases. Effort is also being given to development of the expert knowledge base and the student model, as well as the task selection and task interpretation modules.

The Anticipated Effects of Advanced Mathematics Assessment

The system currently under development is clearly a work in progress and is aimed at developing an environment in which the earlier-discussed ideas about improving assessment can be explored, tested, and developed further. However, it is appropriate to present it in this form so that the revised view of assessment can be discussed and its implications considered. In this section, we summarize the effects that we anticipate our system to have and the kinds of new instructional tools that might develop out of it.

The assessment system envisaged has the potential to contribute more directly to the instructional process. Batteries of tests and test-like questions have always been part of the instructional strategy. The refined knowledge bases, knowledge structures, and student models can be integrated into individual curricula or combined with a computer-based instructional model to help the teacher and student select appropriate sequences of problems.

Habits of the mind

The skills of the expert mathematician probably depend upon a repertoire of invisible mental procedures that, following William James and others, may be called habits of the mind. These are the "important operations which we can perform without thinking of them" mentioned at the beginning of this chapter. In order to acquire a habit, it is important to have a large number of invariant trials. Most classroom systems do not have the resources to arrange and monitor the needed number of invariant trials in mathematical problems. A computer-based system can be used as a problem laboratory for either individuals, tutor-student combinations, or student groups working cooperatively. The students can practice until the latency and correctness of procedural response patterns tells us that the needed habits of the mind have been developed.

Tailored instruction

The student and aggregated class models can be combined to make suggestions to the teacher during preparation of lessons for individual students and for the class. We envision that this would cut down the information-processing load of the teacher and enable the teacher to combine her intuitions, knowledge, and experience to quickly design and prepare superior instructional sequences.

Detailed information

The system will possess and be able to display an exhaustive history of the work of a student or of the work of all the students who have attempted a particular problem. The difficulty with detailed information is that it usually overwhelms one unless, like road maps of increasing detail, one can move easily from one level of detail to the other. With the ability to zoom between levels of detail, access to the detail of how an operation has been attempted can bear rich dividends for both research and instructional design. However, design of tools for the management of such information remains a challenge.

Aggregated information

Different people need aggregated information for various groups of students: subgroups within classes, classes, and larger groups of students (e.g., age cohorts, schools, school districts, states). The math supervisor considering a teacher workshop, a committee of teachers considering a change of textbook, a principal planning for the coming year, a state legislative committee considering a funding formula to focus more resources on math education—all these people and others urgently need better information on aggregated student performance as input to their thinking and decisions.

Organized information

Of all professionals, those in the discipline of mathematics can appreciate the importance of representation systems. Even the nonmathematician can appreciate the advantages of the decimal system, and the undergraduate electrical engineering student can appreciate the power of complex functions. A computer-based assessment system has the potential to build significantly more useful representations of student performance. At ETS, we have been experimenting with various representations of student achievement that can enable one to grasp, at a glance, the level of achievement of a student or group of students at various levels of detail of a domain of mathematics. For example, we may want to see the number of students who understand a selected set of concepts (e.g., limit, function, differential). Or we may want a display of the percentage of the concepts and operations in algebra that have been attained by a student. A user-friendly instructional information system should improve educational decisions enough to be worth the significant effort it will take to create.

Hyper-curricula

Interdisciplinary curriculum efforts appear with regularity, but do not seem to make a significant dent in the traditional subject-centered structure. Intuitively and through persuasive anecdotes, we want to illuminate an idea by problems from other disciplines (e.g., gradient by contour maps from geography). Conversely, we want students to be able to apply their knowledge and skills to novel problems. Finally, we would like to encourage students to build a knowledge structure that has strong links among related concepts. These links (interdisciplinary) should cross disciplinary boundaries without weakening the strength of the intradisciplinary knowledge. An assessment system that generates a broad array of interdisciplinary problems may serve as a tool that will finally lead to the development of successful interdisciplinary instruction.

Conclusion

Assessment serves many functions in our

educational system, but present tests have important limitations. The computer is a very general tool for constructing environments which may improve the quality of assessment and the kind of information available to students, teachers, parents, society and school personnel. We should invest in a serious effort to design and develop computer-based assessment systems that reflect the educational goals we value.

References

Birenbaum, M., & Tatsuoka, K.K. (1987). Open-ended versus multiple-choice response formats It does make a difference for diagnostic purposes. *Applied Psychological Measurement, 11*, 385-395.

Brown, J.S., & Burton, R.B. (1978). Diagnostic models for procedural bugs in basic mathematical skills. *Cognitive Science, 2*, 155-192.

Brown, J.S., & VanLehn, K. (1980). Repair theory: A generative theory of bugs in procedure skills. *Cognitive Science, 4*, 379-426.

Bunderson, R. V., & Forehand, G. (1989). ETS internal document. Princeton, NJ: Educational Testing Service.

Bunderson, C., Inouye, D.K., & Olsen, J.B. (1989). The four generations of computerized educational measurement. In R.L. Linn (Ed.), *Educational Measurement, 3rd edition*. New York: American Council on Education/Macmillan.

Davis, R.B. (1984). *Learning mathematics: The cognitive science approach to mathematics education*. Norwood, NJ: Ablex.

di Sessa, A.A. (1982, January-March). Unlearning aristotelian physics: A study of knowledge-based learning. *Cognitive Science, 6*, (1), 37-75.

Erlwanger, S.H. (1973, Autumn). Benny's conception of rules and answers in IPI mathematics. *Journal of Children's Mathematics Behavior, 1*, (2), 7-26.

Frederiksen, N. (1984, March). The real test bias: influences of testing on teaching and learning, *American Psychologist*, 193-202.

Gitomer, D.H., & Yamamoto, K. (1989, April). *Using embedded cognitive task analysis in assessment*. Paper presented at the annual meeting of the American Educational Research Association, San Francisco.

Hirsch, E. D., Jr. (1987). *Cultural literacy: What every American needs to know*. Boston: Houghton Mifflin.

Hoffman, B. (1962). *The tyranny of testing*. New York: Crowell-Collier.

Holden, C. (1989, May). Computers make slow progress in class. *Science, 244*, 906.

Lipson, J. (1988). Testing in the service of learning: Learning assessment systems that promote educational excellence and equality. *Assessment in the service of learning*. Proceedings of the 1987 invitational conference. Princeton, NJ: Educational Testing Service.

Lundeberge, M. A., & Fox, P. W. (1989, March). *Integrating laboratory and classroom findings on test expectancy*. Paper presented at the annual meeting of the American Educational Research Association, San Francisco, CA.

Masters, G.N., & Mislevy, R.J. (1988). New views of student learning: implications for educational measurement. Unpublished memo.

McCloskey, M., Caramazza, A., & Green, B. (1980). Curvilinear motion in the absence of external forces: Naive beliefs about the motion of objects. *Science, 210*, 1139-1141.

National Council of Teachers of Mathematics (1989). *Curriculum and evaluation standards for school mathematics*. Reston, VA: NCTM.

Rosnick, P., & Clement, J. (1980, Autumn). Learning without understanding: The effect of tutoring strategies on algebra misconceptions. *Journal of Mathematical Behavior, 3*, (1), 3-27.

Schmitt, A.P., & Crocker, L. (1981, April). *Improving examinee performance on multiple-choice tests*. Paper presented at the annual meeting of the American Educational Research Association, Los Angeles.

Snow, R.E. (1987). Aptitude complexes. In R.E. Snow & M.J. Farr (Eds.) *Aptitude learning and instruction, vol. 3*. Hillside, NJ: Lawrence Erlbaum Associates.

Snow, R.E., & Peterson, P.L. (1985). Cognitive analyses of tests: Implications for redesign. In S. E. Embretson (Ed.), *Test design: Developments in psychology and psychometrics*. New York: Academic Press.

Tatsuoka, K.K. (1983, Winter). Rule space: An approach for dealing with misconceptions based on item response theory. *Journal of Educational Measurement, 20*, (4), 345-354.

University of Chicago School Mathematics Project (1989 and 1990). Z. Usiskin & S.L. Senk (Series Directors). Glenview, IL: Scott, Foresman.

VanLehn, K. (1982, Summer). Bugs are not enough: Empirical studies of bugs, impasses, and repairs in procedure skills. *Journal of Mathematical Behavior, 3*, (2), 3-71.

Wainer, H., & Kiely, G.L. (1987). Item clusters and computerized adaptive tests: A case for testlets. *Journal of Educational Measurement, 24*, (3), 185-201.

Wainer, H., Wadkins, J.R.J., & Rogers, A. (1983). *Was there one distractor too many?* Program Statistics Research Technical Report No. 83-

39. Princeton, NJ: Educational Testing Service.

Wise, S. L., & Plake, B. S. (1989). Research on the effects of administering tests via computers. *Educational measurement: Issues and practice, 8*, 5-10.

Yamamoto, K. (1987). A model that combines IRT and latent class models. Unpublished doctoral dissertation, University of Illinois, Champaign-Urbana.

III

Research and Development in Mathematics Assessment

9

Students' Theories About Mathematics and Their Mathematical Knowledge: Multiple Dimensions of Assessment

JOHN G. NICHOLLS, PAUL COBB, ERNA YACKEL,
TERRY WOOD, and GRAYSON WHEATLEY

The use of conventional standardized academic achievement tests to evaluate educational practices and the progress of individual students has long been accompanied by a chorus of dissenting academics—who often stand far off stage—and by muffled grumbling in the ranks of teachers who are constrained to administer these instruments to their charges. But the psychometric juggernaut rolls on, endowed by the arcane language of validity and reliability with an aura of scientific infallibility. The tests are called objective. The scores appear more substantial than any alternative. Opposing these hard and shining instruments, the carping critics offer their own, students', or teachers' subjective impressions.

In one sense, standardized achievement tests do help us avoid some of the problems that can spring from subjective self-interest. Binet's intelligence test, from which current academic achievement tests derive, was constructed to deal with the fact that teachers would sometimes attempt to remove able but troublesome students from their classes by declaring them to be in need of remedial help.

An independently administered ability or achievement test can avoid this form of bias. But technical innovations such as tests cannot save us from bias. Teaching to the test, for example, can create a blatantly false impression of students' mathematical competence. Whatever test is used to evaluate teaching, it almost inexorably biases teaching toward that test (Fredericksen, 1984; Johnston, 1989). The result can be proficiency with the forms of tasks used on the tests but little deep understanding or appreciation of the knowledge that the test constructors hoped to assess. Tests are never objective in the sense of being untarnished by human concerns. Their impact, their usefulness, and their validity depend on how they are construed by humans. Human purposes are, in turn, shaped by the tests and the ways they are used.

For a teacher who seeks to foster students' higher order mathematical thinking, no conventional achievement test is likely to be of immediate help. What teachers can use for this enterprise is more-or-less constant feedback on how students interpret the problems before them. Standardized achievement and aptitude tests (and research instruments including those we describe here) are of no value for this purpose. By the time such measures are scored, any moment of relevance that might be claimed for them is gone. Furthermore, as Johnston (1989) and others have argued, traditional testing practices can undermine the very processes of teaching that many of us want to foster. Formal evaluations cause students who doubt their abilities to perform worse than they are capable of. They also make students reluctant to admit to weaknesses.

The rationale that "to teach you, I have to use these tests to find what you know" is probably the best justification a teacher can come up with for administering such tests. This rationale is not only a half truth, but it also negates the possibility that students might play a more active role in letting the teacher know what does and does not puzzle them. It makes even more remote the prospect that mathematics instruction might help students develop the ability to evaluate their own thinking.

In a variety of ways, traditional tests undermine the atmosphere of mutual trust and joint commitment to the construction of mathematical meaning that makes continuous assessment of student reasoning possible and that makes this assessment an integral, contributing part of learning. Assessment in this sense is no more separable from teaching than the attempt at mutual understanding is separable from an ongoing conversation among friends. If curriculum materials are useful in this conception of education, they are useful in part because they also enable on-going assessment. Conversely, if assessments are useful, it is because they allow teaching to continue. In this framework, the validity of any assessment of learning can only be judged in terms of the negotiated purposes of teachers and students and their theories as to what they are about.

But this is true for all assessments. Validity of assessment always involves matters of construct validity—questions about the meaning of measures. And when one constructs meanings, one also constructs values (Cherryholmes, 1988). The notion of higher order mathematical thinking implies that this type of thinking is, in some sense, more valu-

able than lower order thinking. Why measure something of no importance to us? We do not study everything—we cannot. What we do study reflects what is important to us. What is important is a matter of political and ethical values. Our claim is that teaching and assessing should resemble a conversation among friends intent on making experience meaningful involves a value position. Our judgment that conventional achievement tests are of little value to the teacher trying to promote the shared construction of higher order mathematical insights is also a judgment about the validity of these tests. This judgment reflects beliefs about what sort of teaching is desirable as well as more technical concerns about how to achieve and recognize that teaching and its effects. In short, any test might be valid for some purposes but not for others. Just as a hammer works for some purposes but not for others, so any form of assessment is potentially useful for some but not other political, ethical, and educational purposes. So we want the right tool for our purposes, and we want it to be in good condition.

The measures discussed here are not likely to be of direct use for the practicing teacher. If used to evaluate individual teachers, the measures would probably undermine the ethos that we seek to create. They are, nevertheless, intended to be useful to researchers who want to find out if different educational approaches foster different orientations to mathematics and different levels of mathematical understanding. The greater part of this chapter concerns orientations to mathematics—what is commonly termed motivation, attitudes, and beliefs about mathematics. We submit that these notions can usefully be subsumed under the concept of students' theories about mathematics.

For our purposes, a narrow focus in assessing consequences of mathematics instruction is dangerous. Consider, for example, a study by Helmke, Schneider, and Weinert (1986). They found that, over two years, a form of "direct teaching" produced appreciable gains in achievement on standardized tests at the cost of detrimental attitudes toward mathematics. If gains of one type can involve losses of another, both types of instructional consequence need to be considered. By the same logic, use of only one index of motivation can be problematic. It is common, for example, for researchers to treat perceived ability or expectations of "success" as *the* index of academic motivation. But it is possible for students with high perceived ability in mathematics to lack a desire to improve their understanding.

It follows that to assess validly the consequences of teaching, we should consider as many dimensions of mathematical learning and students' theories about mathematics as possible. But we cannot assess everything. We must select, and this means we cannot make value-free assessments. We should acknowledge this while we gather information that will be relevant to our own educational goals, including information that will expose any undesirable consequences of the pursuit of the things we value. In short, a valid assessment should help us think again about what we are trying to do. It should help us scrutinize our own values and their concomitants. All this is part of the business of validation. It is no business for people who seek a form of scientific validation that is independent of human subjectivity and values.

Students' Theories About School Mathematics

The theme we have advanced with respect to test validity applies to theories of science and to students' theories about academic work, including mathematics. There was a time when, as William James (1907) put it, a good scientific theory was thought of as "absolutely a transcript of reality" (p. 57). Similar is the view, decried by T. S. Kuhn (1970), that science develops as a more-or-less linear accumulation of facts, theories, and methods which will eventually lead us to an error-free understanding of nature. In its use of the metaphor of the lay individual as a naive scientist, attribution theory (Kelley, 1973) implicitly espouses a similar conception of scientific explanation. This is evident, for example, in the research designed to distinguish the effects of biases stemming from personal goals or concerns from veridical or logical attributional processes. In the same vein, Weiner's (1979) attribution theory approach to achievement motivation implies that there is but one basic

(but undefined) form of success (or reality) which different children explain differently. When children report different beliefs about the causes of success (or failure), this is not assumed to reflect different realities or different definitions of success.

An alternative approach to students' theories about classroom events is suggested by pragmatist and Kuhnian conceptions of scientific inquiry (Barnes, 1977; James, 1907; Kuhn, 1970; Rorty, 1983; Toulmin, 1983). In these conceptions, "ideas become...true just insofar as they help us to get into satisfactory relations with other parts of our experience" (James, 1907, p. 58) and any scientific analysis "may from some point of view be useful" (p. 57). There is, in other words, no all-encompassing, value-free position from which one can judge the adequacy of scientific explanations. This does not mean that any explanation will do; it means that what will do depends on what we want done. Scientists vary, not merely in their explanations of any given phenomenon, but in the questions or concerns they have about the phenomenon. They approach any topic with particular interests or concerns. They selectively collect data that are relevant to these concerns, interpret these data in the light of the same concerns, and reconstruct their concerns in light of their interpretations. Scientific controversy is, in this analysis, commonly controversy about what question to ask or how to frame the question. It is not merely controversy about how to answer unambiguous questions.

This perspective resembles the intentional perspective on thought and action (Dennett, 1978; Nicholls, 1989) and the ecological approach to social cognition (McArthur & Baron, 1983) wherein thought and action are interpreted in terms of the thinker's purpose. Thus, rather than judging a causal attribution as veridical or biased according to some absolute external criterion, these theorists argue that bias is inevitable in that selective attention is essential for any meaningful interpretation. Our interpretations can only be useful to us to the extent that they reflect our concerns, and we cannot be concerned about every possible aspect of any situation. All this suggests that different students' explanations for success in mathematics might refer to qualitatively different types of definitions of success or academic goals (Frieze, Francis, & Hanusa, 1983; Maehr & Nicholls, 1980) as well as contribute to the reconstruction of these goals. It might, then, be useful to construe students as scientists—in the post-modern sense—with respect to their personal goals and interpretations of the nature of school mathematics.

Many avenues to the description and assessment of students' theories are possible. The framework described here started with theory on achievement motivation and constructivist and progressive approaches to education (Nicholls, 1989). Interestingly, there is some correspondence between the aspects of student theories we consider desirable and the outcomes advocated by the National Council for Teachers of Mathematics (NCTM) (1989). Thus, the scales described here should, if they are technically adequate, be of value to those whose goals for mathematics education are in accord with those of the NCTM and, more generally, progressive and constructivist philosophies. However, goals and beliefs that others might value are also included.

Several researchers have made a start with two goal dimensions proposed some time ago—task orientation and ego orientation—that are relevant to learning in school (Asch, 1952; Ausubel, Novak, & Hanesian, 1978; Crutchfield, 1962; Maehr & Braskamp, 1986; Nicholls, 1989). These dimensions of individual differences embody different personal definitions of success. Task orientation involves a self-referenced definition of success as the gaining of insight or skill or accomplishing something that is personally challenging. Ego orientation, on the other hand, means that to experience success, the student must establish his or her ability as superior to that of others. In this usage, "ego" has the meaning illustrated by Einstein's assertion that "the desire to be acknowledged as better, stronger, or more intelligent than a fellow...leads to an excessively egotistic psychological adjustment" (1956, p. 34). Ego orientation means that although one might seek to learn, to gain insight, or to perform one's best, these actions would not be ends in themselves, but rather means to the end of establishing one's ability as superior.

Studies of adults (Maehr & Braskamp, 1986), high school students (Nicholls, Patash-

nick, & Nolen, 1985), and junior high students (Nolen, 1988; Thorkildsen, 1988) indicate that the two goal dimensions of task and ego orientation are virtually independent of one another. There is not, as some would have it, a single bipolar dimension of task versus ego orientation. A third orientation—work avoidance—is generally negatively related to task orientation and unrelated or positively related to ego orientation (Nicholls et al., 1985; Thorkildsen, 1988).

It was expected that, for ego-oriented students, the concept of ability would be very important, and success (as they define it) would be seen as dependent on superior ability. To task-oriented students, however, concepts of learning, understanding, and effort should be important and, therefore, attempts to make sense of mathematics should appear essential for success. Evidence supporting these predictions has been obtained with older students and with reference to school in general (Nicholls, 1989; Nicholls et al., 1989; Thorkildsen, 1988).

We adapted the motivational orientation (i.e., personal goals) scales to make them appropriate for second grade mathematics students (Nicholls et al., in press). The scales used to assess beliefs about the causes of success (i.e., what students must do to do well) were similarly adapted. Four scales assessed beliefs that success in mathematics depends on (i) effort, (ii) attempts to understand mathematics, (iii) superior ability and attempts to beat others while contemplating mathematical tasks, and (iv) task-extrinsic forms of conduct, such as being quiet during mathematics instruction.

In this second grade sample, task orientation and ego orientation in mathematics were evident as two individual difference dimensions. Students' status on one dimension was not a reliable indicator of their status on the other. Work avoidance (i.e., desire to do as little as possible) was negatively associated with task orientation and positively associated with ego orientation. Task orientation was moderately associated with the beliefs that success depends on interest, effort, attempts to make sense of things, and collaboration with one's peers. Ego orientation was not associated with these beliefs, but was moderately highly associated with the beliefs that success depends on superior ability and attempts to beat one's peers. In other words, by the second grade (if not before), moderately internally consistent or rational complexes of academic goals and beliefs about how success in school mathematics is achieved are evident (Nicholls et al., in press).

Program effects

The preceding evidence on dimensions of individual differences in students' theories of mathematics provided us with an initial indication of the validity of the scales for assessing students' orientations to school mathematics. Next, we compared five conventional classrooms with a second grade classroom where mathematics teaching was consistent with the constructivist tradition. The scales were administered to these six classes at the end of the school year. Students from a single school had been randomly assigned to these classes at the beginning of the year. Only instruction in the target class was compatible with implications derived from a constructivist, progressive perspective (cf. Cobb, 1986, 1988; Confrey, 1986; DeVries & Kohlberg, 1987; Dewey, 1966; Kohlberg & Mayer, 1972; von Glaserfeld, 1983). This class also embodied the NCTM recommendation that the

... emphasis should be on establishing a climate that places critical thinking at the heart of instruction. All statements [concerning mathematics] should be subject to question, and both teacher and children should be open to reaction and elaboration from others in the classroom. Discussing and validating children's thinking and solutions should be a major instructional activity for both students and the teacher (p. 26).

The mathematics lessons in the target class were marked by an atmosphere of dialogue and collaborative problem solving (Cobb, Wood, & Yackel, in press a). After instructional activities were introduced, students worked cooperatively in pairs and then participated in teacher-orchestrated, whole-class discussions of their interpretations and solutions. In contrast to traditional mathematics instruction, the teacher did not attempt to steer the students to pre-conceived solutions (Voigt, in press), but attempted to

facilitate dialogue among students (Cobb, Wood, & Yackel, in press a; Wood & Yackel, in press). When, as frequently happened, students offered two or more conflicting solutions, the teacher typically framed the situation as a problem for the students to resolve by justifying and explaining their solutions.

The problem-centered emphasis was facilitated by instructional activities that were designed to give rise to personally challenging mathematical problems for second graders at a variety of different conceptual levels. With regard to the instructional activities in arithmetic, for example, initial activities involving addition and subtraction were designed to facilitate the construction of thinking or derived fact strategies and increasingly sophisticated concepts of addition and subtraction. Subsequent activities were designed to encourage the development of solution procedures involving composite whole-number units (i.e., numerical units themselves composed of ones). These activities included initial multiplicative and divisional situations and tasks involving money as well as activities that give rise to discussions that reflect the children's conceptions of numeration. The activities typically used in the latter half of the school year gave further opportunities to construct increasingly sophisticated concepts of ten while completing tasks involving the addition and subtraction of two-digit numbers. Multiplicative and divisional situations were also extended and elaborated in this part of the school year. In addition to symbolic sentences, activities used throughout the year included the full range of arithmetic word problems identified by research-based classification schemes (Carpenter & Moser, 1984; Kouba, 1989), activities involving spatial dot patterns, and balance activities (Figure 9.1). Sample thinking strategy activities, together with excerpts of both small group and whole class dialogues, are presented and discussed by Cobb and Merkel (1989).

The interweaving of additive and subtractive with multiplicative and divisional situations reflects the contention that children's construction of increasingly sophisticated concepts of 10 and of multiplicative units can be accounted for by the construction of a single sequence of conceptual operations

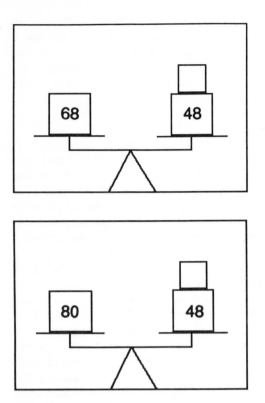
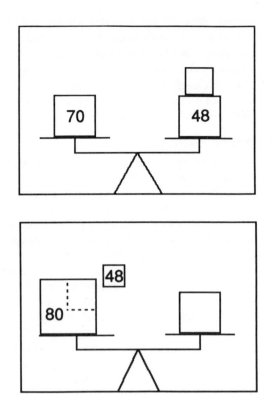

Figure 9.1. Sample balance activity sheet.

(Steffe, 1989; Steffe & Cobb, 1984). One of the primary goals when developing the instructional activities was to facilitate children's construction of these conceptual operations in a variety of computational situations.

Case studies of the target classroom (Cobb, Wood, & Yackel, in press b; Cobb, Yackel, & Wood, 1989) document the processes by which the teacher initiated and guided the renegotiation of classroom social norms and thus influenced students' implicit theories about mathematics instruction. As part of this process, she attempted to help the children develop atypical obligations during mathematics instruction and, in turn, accepted nontraditional obligations for her own activity. These included taking students' mathematical interpretations seriously as she subtly initiated and guided the negotiation of mathematical meanings without making overt evaluations of students' contributions. Other contrasts with traditional instruction included the lack of grading of written work and the complete absence of individual paper-and-pencil seatwork. In discussing her experiences during the year, the teacher clearly indicated

her feeling that the ethos of her class was very different both in the way it compared to her classes of previous years and to the ethos in her colleagues' classrooms.

Convergence of the case study and the teacher's impressions with the picture produced by our scales would help validate each type of analysis. As a result of its distinctive characteristics, the target class was expected to score higher than others on task orientation and the belief that success in mathematics depends on attempts to make sense. The class was also expected to measure lower on work avoidance, ego orientation, and belief in the importance of task-extrinsic factors for success. The differences between the target class and the others were consistent with these expectations (Nicholls et al., in press), suggesting that the motivational goals of the mathematics program had been achieved and, reciprocally, that the scales serve reasonably well the purpose we hoped they would.

An additional interesting finding was that some scales that were quite strongly associated at the level of individual differences showed divergent trends in the comparison of classes. First, for individuals, task orientation was associated with — and loaded on the same factor as — the belief that success in mathematics depends on attempting to understand. The project class was matched by only one other class on task orientation. Yet, this other class was significantly lower than the project class (and lowest of the six classes) in the belief that success depends on trying to make sense of mathematics.

Second, although the target students were more certain than those in other classes that success in mathematics would be promoted by liking to think, trying to figure things out, and trying to understand instead of just getting answers, the classes did not differ significantly in beliefs about the importance of interest and effort to success in mathematics. Many teachers and some research perspectives on student motivation emphasize that students should adopt the view that effort will cause academic success, and some advocate training students who lack motivation to attribute their failures to lack of effort. The constructivist and progressive perspectives are distinguished from such approaches by a concern that effort be directed specifically at the making of meaning. The result described in this section suggests that our scales are sensitive to this critical distinction. Even young students can distinguish mere effort from the deployment of effort specifically to gain mathematical insight — even though this distinction was not apparent in the individual difference analyses wherein the belief that success in mathematics requires effort loaded on the same factor as the belief that success requires attempts to understand (Nicholls et al., in press).

These discrepancies in results for individual students and for classes suggest the importance in program evaluation of maintaining distinctions such as those made by our subscales concerning students theories — even when there is not strong evidence for this at the individual level.

These results also suggest the advantages of a multidimensional collection of scales. If one had employed any one of these motivational indices (or, for example, simply a measure of liking for or of intrinsic motivation in mathematics), an appreciably less-informative picture of the motivational climate of the target class would have been obtained. The fact that the picture obtained with the scales converged with that reported by the teacher of the target class and those who closely observed it strengthens the claims of each of these assessments.

Related technical and ethical considerations in the assessment of academic motivation

It is common for researchers to emphasize perceived competence or self-concept of ability as an important index of academic motivation. It is very easy to gather evidence that higher perceived ability is associated with higher attainment and a variety of indices of adjustment (Taylor & Brown, 1988). But it does not follow that high perceived ability in mathematics should be a program goal or a goal for individual students.

Generally overlooked is the basis of students' responses to self-concept of ability scales. Even when these scales do not explicitly refer to one's standing relative to others, students interpret them in these terms. Thus, "I am good at mathematics" is construed as "I am better at mathematics than most others" (Nicholls et al., 1989). This is a very reasonable interpretation on the part of anyone presented with such a question. But as a general goal for instruction, high perceived ability in this sense is of questionable value. Everyone can't be above average. A little unrealistic optimism about one's abilities might be helpful now and then, but there must be some limit.

The preoccupation on the part of researchers with assessing students' perceptions of their ability might be seen as one expression of our society's tendency to define success in terms of "victory over your competitors" (Lasch, 1978, p. 58). We might reflect on the fact that Chinese students, who some would have us imitate, have lower perceived competence than students from the United States (Stigler, Smith, & Mao, 1985).

A desire to make sense and engage in meaningful activity is not equivalent to a desire to be superior to others. As we have shown, they are more-or-less uncorrelated. Furthermore, high perceived ability or attainment is not appreciably associated with task orientation or the belief that the way to succeed at mathematics is to collaborate with others in making meaning. This was one additional result of the study we just described (Nicholls et al., in press). It parallels results of studies of orientation to school in general (Nicholls, 1989; Nicholls et al., 1989) and to science (Nolen, 1988). Furthermore, the project class in the described study was not outstanding in perceived ability. It is not necessary, then, to make students believe that they are above average to promote a constructive – and constructivist – orientation to learning.

A number of researchers find perceived ability positively associated with intrinsic motivation – a result that would cast doubt on the immediately preceding conclusion. But there might be technical reasons for this. Constructors and users of questionnaires can easily overlook occasions when different scales have nearly equivalent if not equivalent items (Nicholls, Licht, & Pearl, 1982). Thus, they often construct elaborate explanations for associations that might have a more mundane explanation. Namely, that if you ask people the same question twice, they tend to give the same answer each time. This applies to some self-concept and some intrinsic motivation questionnaires. Anyone contemplating using a perceived ability scale should check that it does not have items asking students whether they like school work. It should ask only how able they think they are. By the same token, a measure of intrinsic motivation should not ask students whether they are able or like hard academic tasks. A hard task is one few students can do, and, other things being equal, students with higher perceived ability will be more likely to prefer such tasks. It may not follow that they are more inclined to seek understanding.

Misleading results occur when scales that are presented as measuring some aspect of motivation in fact contain direct or indirect references to perceived ability. If assessed adequately, perceived ability is, at least in older elementary school students, highly associated with indices of achievement or learning (Nicholls et al., 1989). The simplest interpretation of this finding is that when you ask someone how able they are, they can tell you fairly accurately how they stand. This is not, in itself, of great educational significance. Yet, results that reflect this phenomenon are

often presented as establishing that a measure of motivation is a significant index (whether cause or effect of achievement) of motivation. On the contrary, high associations between a measure of motivation and academic attainment should make one suspect that this relationship might merely reflect students' knowledge of their standing and, therefore, be educationally inconsequential. Thus, for the work presented in this chapter, it is important to note that perceived ability is not appreciably associated with the goals and beliefs scales. In the same vein, we note evidence that task orientation predicts use of deep processing strategies for reading passages on science when perceived ability is partialled out (Nolen, 1988).

None of this is to say that perceived ability is unimportant in student motivation. It often is very important. The attempt to avoid feelings of incompetence can occasion many unfortunate behaviors (Covington, 1984; Covington & Beery, 1976). But it remains a mistake to leap from such observations to the conclusion that enhancing perceived ability is the answer. More useful, we suggest, is the attempt to create classrooms where students see success as dependent on collaborative attempts to construct meaning rather than on superior ability.

Refinements and Additions: A Second Study

In 1988, the mathematics research and development project discussed in the preceding section was expanded to include a larger number of teachers. This offered the opportunity to improve and add to our scales, to examine further their validity, and to learn more about the program. Measures of mathematics learning were also added in this phase of the project. These were of interest in their own right and were a source of further evidence concerning the construct validity of the measures of student theories. In turn, the measures of goals and beliefs provided data relevant to the validity of the measures of learning.

Dimensions of students' theories revisited

The main change made in the measures of

goals and beliefs about school mathematics involved the addition of items concerning collaboration with fellow students. Previous work showed task orientation (but not the other orientations) to be positively correlated with the belief that success depends on collaboration with peers. This belief was also associated with the beliefs that success in school depends on interest and attempts to understand (Nicholls, 1989; Thorkildsen, 1988; Nicholls et al., in press). Furthermore, task orientation (more than the other orientations) tends to be associated with the view that school should help prepare one for adult work that will contribute to the lives of others (Nicholls, 1989; Thorkildsen, 1988). That is, task orientation and its associated beliefs appear to be consistent with a tendency to see oneself as having social responsibilities and to see others as resources for learning.

It is interesting that these results obtained even in typical school situations where, as is the norm, children had little experience of collaborative learning. Presumably, the task-oriented students' tendency to see collaboration as helping promote success is not a result of the school's emphasis on collaboration. This result might reflect children's implicit, if not explicit, recognition of the importance of social transaction in the construction of knowledge. Piaget's later work conveyed the impression of children as solitary intellectual voyagers, responsive only to contradictions between their own schemes and between the way things appeared and the way they expected them to appear (Furth, 1981). The importance of peers in stimulating thought and in modeling effective intellectual adaptations has been accorded more recognition recently (Damon, 1981; Doise & Mugny, 1984; Sigel, 1981). Children might have recognized this all along, as indicated by evidence that task-oriented students, who see success in mathematics as requiring attempts to make sense, also see collaboration as likely to help promote success as they define it. It is also relevant that children see peer tutoring (which rarely occurs in school) as the fairest way of dealing with the fact that some students learn more rapidly than others (Thorkildsen, 1989). Thus, a classroom emphasizing understanding and collaboration might also be seen by students as a just classroom.

Collaboration among students was a central feature of the present project. It thus reflected progressive views to education (Dewey, 1966; Nicholls, 1989) and the arguments of many who advocate cooperative learning (Johnson, Johnson, & Scott, 1978; Sharan et al., 1984; Yackel, Cobb, & Wood, in press). These views can be contrasted with the recommendation of some writers who would foster students' perceptions of high ability by limiting their opportunities to compare their understanding with that of others (Rosenholtz & Simpson, 1984). By way of contrast, our view is that, on any task, there are always individual differences in competence or knowledge, and that there is nothing inherently wrong if students recognize their standing relative to their peers. Problems occur only if superiority over others becomes a controlling end of learning. But it does not follow that we must avoid these problems by having students engage in activities that prevent them from comparing their solutions to problems. This seems to involve an extreme and unnecessary individualism. Conflict of ideas—without a conflict of egos—is the essence of democratic, progressive, constructivist education and life.

In previous work, the belief in the importance of collaboration to success in mathematics had been assessed with only one item. To make the interpretation of results less equivocal, more items (worded differently and assessing different aspects of collaboration) were added. In addition, items assessing the personal goal of collaborating in mathematics learning were added.

When the new motivational orientation items were factored and item-analyzed, the results accorded with our expectations. For example, the following items all loaded on a task orientation factor: "I feel really pleased in math when we help each other figure things out," "...when everyone understands the work," and "...when I find a new way to solve a problem." A parallel result was obtained for beliefs about the causes of success in mathematics: one scale included the beliefs that students succeed in mathematics if they "try to explain their ideas to other students," "try to understand each other's ideas," and "don't give up on really hard problems." The revised motivational orientation (i.e., students' personal goals in mathematics) scales are shown in Table 9.1, and the scales assessing beliefs about the causes of success in mathematics (beliefs about the "reality" of school mathematics) are shown in Table 9.2.

A further aspect of students' theories about school mathematics was explored in our second study—beliefs about the consequences of learning mathematics in school. We emphasized long-term influences of mathematics education that are consistent with the educational goals articulated above, and, to provide a clear contrast, a set of outcomes that, at least on the face of it, appeared wildly irrelevant to mathematics and only likely to be endorsed by anyone with a jaundiced view of the relevance and usefulness of mathematics education. Factor analysis produced two factors, which were used as the basis for forming the two scales shown in Table 9.3. The first embodies the view that learning mathematics improves students' ability to explain themselves, figure things out, know if things make sense, and understand others' ideas. The second includes the beliefs that mathematics increases students' running ability, ability to memorize long lists, makes them do what others say, and makes them want to watch lots of TV when they grow up.

When scores on the scales assessing motivational orientations, beliefs about the causes of success, and beliefs about the consequences of learning mathematics were factor analyzed together, similar factors to those obtained in the first study (Nicholls et al., in press) were obtained (see Table 9.4). Factors dominated by task orientation and ego orientation were again evident. The task orientation factor involved the goals of working hard, figuring things out, and collaborating; the beliefs that success requires effort, interest, collaboration, and attempts to understand; and the beliefs that learning mathematics improves people's ability to make sense of things and to explain themselves. Even at the second grade level, the views that mathematics learning helps one think independently, understand, and make sense of the world are linked conceptually to related beliefs about how people can most successfully "do" mathematics and to students' personal criteria of success in mathematics.

This result supports the earlier evidence that second graders exhibit conceptually

Table 9.1. Motivational orientation (personal goal) scales.[a]

Task orientation I: Effort (alpha = .78)

5.	I solve a problem by working hard.
8.	the problems make me think hard.
9.	what the teacher says makes me think.
10.	I keep busy.
11.	I work hard all the time.

Task orientation II: Understanding and collaboration (alpha = .71)

3.	something I learn makes me want to find out more.
4.	I find a new way to solve a problem.
6.	something I figure out really makes sense.
20.	everyone understands the work.
21.	we help each other figure things out.
22.	other students understand my ideas.

Ego orientation (alpha = .83)

16.	I know more than the others.
17.	I finish before my friends.
18.	I get more answers right than my friends.
19.	I am the only one who can answer a question.

Work avoidance (alpha = .79)

2.	it is easy to get the answers right.
12.	I don't have to work hard.
13.	all the work is easy.
14.	the teacher doesn't ask hard questions.

[a]The general question for all these items was "What makes you feel really pleased when you are doing math in school?" *Note:* The stem for every item (not included above) was "I feel really pleased in math when..." This with the rest of each item was read aloud by researchers. The response scale was "YES, yes, ?, no, NO." An introductory discussion made the point that different people are made pleased by different things — a notion students readily acknowledge — and that the questionnaire concerned such personal preferences. Examples of types of foods, games, and other events were discussed, with examples elicited from the students to make sure that they understood that the task resembled voting or answering an opinion poll. Confidentiality was assured, with the parallel of the opinion poll being used to make the point that the researchers did not need to know which individuals gave which answers to get the information they needed.

coherent and encompassing theories of mathematics. This conclusion is also supported by findings concerning the scale assessing beliefs that mathematics learning contributes to apparently extraneous desires such as watching

Table 9.2. Scales assessing beliefs about the causes of success in mathematics.[a]

A. Success requires interest and effort (alpha = .71)

11.	they work really hard.
12.	they always do their best.
16.	they like to think about math.
17.	they are interested in learning.

B. Success requires collaboration and attempts to understand (alpha = .62)

14.	they try to explain their ideas to other students.
15.	they try to understand each other's ideas about math.
18.	they try to understand instead of just get answers to problems.
19.	they try to figure things out.
20.	they don't give up on really hard problems.

C. Success requires competitiveness (alpha = .67)

6.	they try to get more things right than the others.
7.	they try to do more work than their friends.
8.	they are smarter than the others.

D. Success requires conformity (alpha = .76)

4.	they solve the problems the way the teacher shows them and don't think up their own ways.
5.	they all solve the problems the same way and don't think up different ways.
9.	they try to find their own ways of doing problems.[b]
10.	they like to find different ways to solve problems.[b]

E. Success reflects extrinsic factors (alpha = .58)

1.	they are just lucky.
2.	their papers are neat.
3.	they are quiet in class.

[a]The general question for these items was "What will help students to do well at math?"   [b]These items were reversed.   *Note:* All items had the stem, "Students will do well in math if…" Each complete item was read aloud. Response format as in Table 9.1. The introductory discussion of these scales stressed how everyone had different opinions or theories about the best way to do things and the "we" were interested in students' theories about how students do well in math, and which sort of students do well in math.

Table 9.3. Scales assessing views about the consequences of learning mathematics.[a]

A. Independence and knowledge (alpha = .61)

1.	makes students good at explaining their own ideas.
3.	helps students think for themselves.
4.	makes students want to do lots of math when they grow up.
6.	helps students learn how to figure things out.
7.	helps students know if something makes sense.
10.	helps students understand other people's ideas.

B. Irrelevant consequences (alpha = .64)

2.	helps students run fast.
5.	makes students do what other people tell them.
8.	makes students want to watch lots of TV when they grow up.
9.	helps students remember long lists of things.

[a]The general question for these items was "What does learning math do to people?"   *Note:* All items had the stem "Learning math…" Other details as for Tables 9.1 and 9.2.

TV and to the abilities to memorize long lists and run fast. This scale loaded on the same factor as the beliefs that success in mathematics depends on conformity and extrinsic factors.

Influences of a problem-centered mathematics program

The sample comprised all 330 second grade students from three schools in one county system in which ten of the 18 classes were taught in a fashion consistent with the program described in the first section of this chapter. Within each school, students had been randomly assigned to classes (Cobb et al., 1989).

The comparison of program and nonprogram students produced results that were different in some details, but overall conveyed the same message as the first study. Students from the project classes were significantly less ego-oriented than other students, but did not differ on the other personal goal scales. As in the first study, the project students did not differ significantly from others in the belief that effort and interest lead to success. (All

Table 9.4. Oblimin (oblique) factors of motivational orientation, beliefs about the causes of success, and views about the consequences of learning mathematics.

	I	II	III
Orientation			
Effort (Task I)	.69	− .40	.31
Understanding	.75		
Ego		.79	
Work avoidance	− .32	.62	
Beliefs/success			
Effort	.80		
Understanding	.73		
Competitiveness		.63	
Conformity			.70
Extrinsic			.68
Effects of math			
Independence	.68		
Extrinsic			.70

Note: Loadings below .30 are not shown. There were three eigenvalues above 1, and the first three factors accounted for 58% of the variance. The highest correlation between any pair of factors was .08. Slightly different results were obtained for girls and boys when they were analyzed separately.

strongly agreed that they do.) But project students asserted more decisively that attempting to understand both collaboratively and individually leads to success. The project students were also less likely to see success as resulting from competitiveness or other extrinsic factors. They decisively rejected the notion that conformity to teacher or peer methods would lead to success, whereas the nonproject students tended to endorse this view. Finally, although the groups were similar in their tendency to view mathematics learning as leading to intellectual autonomy, the project students were more inclined to reject the notions that mathematics learning could make people, for example, fast runners, good memorizers, couch potatoes, and compliant to others' demands.

It is perhaps somewhat surprising that the groups did not differ in the goal of understanding (task orientation II). (Both groups rated this goal highly.) It should be remembered that the motivational orientation scales ask each student whether understanding mathematics, out-performing others, and so on makes him or her feel pleased. These scales do not ask how often students obtain these sources of satisfaction in school. That is, the scales are intended to assess personal academic goals. It is perhaps comforting that project and nonproject students declare that they find it pleasing when they make sense of mathematics and that they are more strongly task-oriented than ego-oriented.

However, our data raise the prospect that this commitment to making sense of mathematics might not hold up under traditional methods. In both studies, the belief that success in mathematics requires collaboration and attempts to understand distinguished the project classes from the others. In project classes, the level of task orientation II (collaboration and understanding) was matched by the level of beliefs that success requires collaboration and attempts to understand. Nonproject students, however, were less likely to see success as dependent on collaboration and trying to understand. In other words, for nonproject students, the reality of the classroom, as they see it, is less closely aligned with their personal goals of insight and collaboration. It would not be surprising if the persistence of such a discrepancy led to lowered commitment to making sense of mathematics and collaboration with peers. It could, in fact, eventually prove to be a source of alienation from mathematics.

Beliefs about causes of success in mathematics, being beliefs about the way things happen in the classroom, are likely to reflect changes in classroom practice more sensitively than are students' personal goals. Yet, changes in these beliefs might be harbingers of change in personal goals. Over how many grades will students retain a commitment to collaboratively making sense of mathematics if they see these things as playing a smaller role than they would wish?

Influence of the program on mathematics learning

To assess the influence of the program on mathematics learning, we developed a new test which focuses on arithmetical computation abilities and related numeration and whole number operations and concepts. We chose this focus for two reasons. First, computation is the area of second grade mathematics that a priori seems the most difficult to teach and assess in a way compatible with recent reform recommendations. For example, a relatively high proportion of the exemplary mathematics activities given in the NCTM standards are drawn from geometry, probability, and statistics—areas that at first blush seem to lend themselves to active exploration more easily than does computation. Second, arithmetical computation is the heart of elementary school mathematics for many constituencies not directly involved in the current reform debate. In an era of accountability testing, we have found it helpful to document children's computational proficiency to gain credibility with parents and school officials who are more remote than teachers from the immediate and ongoing nature of learning in the classroom.

The arithmetic test (Cobb et al., 1989) was comprised of *instrumental* and *relational scales* which corresponded to two factors obtained when the subscales were factor analyzed. The instrumental scale consisted of seven two-digit addition and subtraction tasks in vertical column format and five in horizontal sentence format. Five of the same number combinations were used across formats (e.g., 19 + 54 = _____ and 73 − 28 = ____). This scale is labeled "instrumental" in that it is possible to perform well by using standard computational algorithms without conceptual understanding.

In contrast, items on the relational scale were designed to assess students' conceptual understanding of place value numeration and their computational ability in nontextbook formats. The computational items on the relational scale included four two-digit addition and subtraction tasks presented in everyday language. Two of these tasks involved the same number combinations as tasks presented in vertical column format. One two-digit missing addend task was also posed in the everyday language format and two in the horizontal sentence format, together with one two-digit change word problem with the change unknown (i.e., a two-digit missing addend word problem).

The numeration items of the relational scale included typical textbook tasks such as "How many tens in 28?," together with items such as "What number do 12 ones and 3 tens make?" Three further items were adapted from the work of Steffe et al. (1988) and two from the work of Kamii (1986). An example of the first type is shown in Figure 9.2. An overhead projector was used to present both types of items. The children's test booklets contained pictures that matched the image projected on the screen. With reference to Figure 9.2, the administrator first explained that each strip contained ten of the small squares, and that the rectangular region represented a cloth that hid some more squares. The students were then told, "There are fifteen squares under here (*points to the rectangular region*). We can see these squares (*points to the strips and squares*). How many are there altogether (*circles the entire display with a finger*)?" The question was then repeated for a

There are 15 hidden. How many are there altogether?

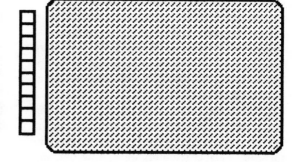

Figure 9.2. Sample relational scale item.

second time. With regard to the items derived from Kamii's work, the administrator first told the students that there were 23 dots in the collection shown in their booklets. Most students counted to check. The administrator then wrote "23" on the overhead. Next, the "2" was covered and the students were asked to "circle some of your dots to help someone understand what this part of the number means (*draws a circle around the "3"*)." When the students had done so, the "2" was uncovered and the "3" covered. The same question was then repeated as the administrator drew a circle around the "2."

Project students were probably more familiar than nonproject students with the horizontal sentence and everyday language formats. However, in contrast to their nonproject peers, the project students had not seen typical textbook numeration tasks and were less familiar with the vertical column format. Further, neither group had previously seen the Steffe or Kamii items. This instrument was administered in May together with the motivation and belief instruments by the project staff.

In addition to the project test, the teachers administered the state-mandated ISTEP test in March. The mathematics portion of this test is composed of two sections — computation, and concepts and applications. The California Achievement Test comprises 80 percent of this test. The competencies assessed are computational facility in familiar textbook settings, identifying typical textbook figurative icons (e.g., three sets of four items to be identified as "$3 \times 4 = 12$") and decoding conventional symbol strings (e.g., which number is in the tens place of 346?). Thus, it is possible to answer most items correctly by mastering the figurative conventions institutionalized in the elementary school mathematics culture. In addition, students were administered the ISTEP cognitive skills index. Responses on this scale indicated that there was not a significant difference in general aptitude between the project and nonproject student populations.

There were no significant differences between project and nonproject students on the ISTEP computation subtest or the instrumental scale. Project students were, however, significantly superior on both the ISTEP concepts and applications subtest and the relational scale. In the case of the relational scale, the difference in group means was approximately one standard deviation.

These above results of arithmetical performance suggest that project students had no special advantage in stereotypical textbook computation settings, but that their numeration concepts were developmentally more sophisticated. Three further analyses that we subsequently conducted indicated that project students more frequently used self-generated methods to solve tasks on the instrumental scale, were less influenced by task format, and were less likely to use figurative rules.

The first of these analyses dealt with the students' use and non-use of standard vertical algorithms. Separate analyses were conducted for each of five groups of computational tasks: the tasks derived from the work of Steffe et al. (1988), the everyday language tasks, the missing addend word problem, the horizontal sentences, and the tasks in column textbook format. The differences in the project and nonproject students' use of vertical algorithms were not significant for the Steffe tasks, everyday language tasks, or the story problem. (Less than 20 percent of students in both the project and nonproject groups used a standard algorithm on any of the three groups of tasks.) In contrast, vertical algorithm use was less frequent for project students on both the horizontal sentences (30 percent versus 44 percent for nonproject students) and the column textbook tasks (31 percent versus 82 percent for nonproject students). This finding indicates that the two groups used similar methods to complete the computational items of the relational scale but not of the instrumental scale, the difference being particularly striking for the tasks in column format. The data also indicate that nonproject students' use of standard algorithms was relatively inflexible. The frequency of use fell dramatically as tasks became increasingly dissimilar to the column format in which use of the algorithms had been taught.

The differences in the two groups of students' computational methods were further investigated by focusing on the seven pairs of addition or subtraction items in which the same number combinations were used in vertical and nonvertical task formats.

Nonproject students found the tasks in vertical format appreciably easier and the results suggest that, in general, the computational methods of nonproject students were less flexible and more format-specific than those of project students.

In a further attempt to gather information about the nature of students' solution methods, it was possible a priori to identify answers to four tasks that indicated that students had followed figurative rules (e.g., "What number does 2 ones and 4 tens make? 24.") Again, the groups differed significantly. Nonproject students used figurative rules almost three times as frequently as project students.

These three follow-up analyses lead one to suspect that the similar scores on both the instrumental scale and the ISTEP computation subtest mask differences in the quality of the two groups of students' computational activity. Project students' eschewal of figurative rules and the greater flexibility of their computational methods indicate that their methods reflect a deeper conceptual understanding. This is, of course, consistent with their superior performance on the relational scale. The general pattern of performance on the arithmetic tests is also consistent with the analyses of the two groups' beliefs about the causes of success in mathematics. Nonproject students' use of vertical algorithms in the familiar textbook format is compatible with their beliefs about conforming to the solution procedures of others rather than developing their own methods. Similarly, project students' superior performance on the relational scale and their limited use of figurative rules adds credibility to their espoused belief in the importance of understanding. By considering a broad range of measures, we begin to glimpse important differences in both the thinking by which project and nonproject students calculated correct answers and the beliefs that support that thinking.

Individual differences in students' theories and mathematical knowledge

As a final check on the validity of the scales assessing the different aspects of students' theories about mathematics, correlations of these scales with the measures of instrumental and relational knowledge were computed.

Table 9.5. Correlations of instrumental and relational knowledge of mathematics with motivational orientations, beliefs about the causes of success, and views about the consequences of mathematics learning.

	Instrumental knowledge	Relational knowledge
Orientations		
Effort (Task I)	$.15^b$	$.13^b$
Understanding (Task II)	$.20^c$	$.17^c$
Ego	$-.03$	$-.01$
Work avoidance	$-.23^c$	$-.26^c$
Beliefs/success		
Effort	$.20^c$	$.17^c$
Understanding	$.23^c$	$.15^c$
Competitiveness	$-.11$	$-.09^a$
Conformity	$-.11^a$	$-.13^b$
Extrinsic	$-.25^c$	$-.24^c$
Effects of math		
Independence	$.08$	$-.04$
Extrinsic	$-.26^c$	$-.31^c$
Multiple correlation	$.41^c$	$.40^c$

Note: Variation across classes was controlled in these correlations by converting all individual scores to z-scores, calculated separately for each class. $^a p < .05$; $^b p < .01$; $^c p < .001$, all tests two-tailed.

Variation across classrooms was controlled for all variables, so that program/nonprogram influences and other consequences of class differences should not contribute to these associations.

As shown in Table 9.5, results for the relational and instrumental knowledge tests were similar. When the scales assessing theories were entered in a step-wise regression predicting performance on the two scales, they accounted for about 16 percent of the variance in both instances. Higher scores on the scales were associated with task orientation, beliefs that success in mathematics requires interest, effort, attempts to understand and collaborate, and with rejection of the view that learning school mathematics will have clearly extraneous or extrinsic consequences. Low scores were associated with work avoidance and the beliefs that success follows from competitiveness, conformity, and extrinsic factors such as luck and "nice" clothes and behaviors.

In summary, those dimensions of students' theories that we deemed constructive generally correlated positively with the learning measures, while the dimensions that appeared not constructive were negatively associated with mathematical learning.

Again, we return to the contrast between project, nonproject, and individual differences. In both studies, the project led to increases in some of the desirable dimensions, while not affecting a second group of dimensions which were, at the level of individual differences, quite highly associated with the former. As shown in Table 9.5, the goal of understanding (task orientation II) and the associated beliefs that success requires collaboration and attempts to understand were both associated with more advanced mathematical performance. But the project classes were superior on only these beliefs, not on task orientation II. Similarly, at the level of individual differences in associations with the different aspects of students' theories, the relational and instrumental scales did not look very different (Table 9.5). But a difference was clear when project and nonproject students were compared.

Conclusion

Conventional achievement tests might seem to produce hard, substantial data. One should not forget, however, that, like questionnaires, they involve marks that students make on paper. Their substantiality, to the extent that it exists, depends on our agreement to construe these marks as important and substantial. If students scored high on tests that are sensitive to higher order thinking but thought mathematics had nothing to do with making life meaningful or interesting, would we have achieved anything of substance or of a high order? What do any of our achievements profit us if they fail to make our lives subjectively meaningful or somehow satisfying? In education, we hope to make learning more meaningful, more connected to the rest of students' lives. This does not mean that we can ignore measures of attainemnt or reasoning.

In the case of the classes we studied, the promotion of insightful learning was accompanied by beliefs that appeared to auger well for mathematics education, for students as individuals, and for the social fabric. The project experience appeared to dispose students to see school mathematics as the meaningful activity most of them want it to be. This makes the higher levels of rational reasoning of the project students appear to be a genuinely substantial achievement. The tendency of the project to lead students to reject the notions that conformity leads to success and that learning mathematics makes one likely to conform to whatever others suggest might also foretell of personal and intellectual autonomy. Finally, for a society often accused of extreme individualism and competitiveness, it appears significant that these strengths – that some might have thought to be the products of competition – were accomplished in an atmosphere of collaboration, with reduced levels of ego orientation and elevated beliefs in the contribution of collaboration to success.

These results, based on marks made on paper by second graders, converge with experiences of teachers and observers and with the theoretical frameworks within which the measures were conceived. These frameworks, in turn, are nothing more lofty than attempts to make sense of the experiences of being students and teachers. Our measures are somewhat more formal, but, in turn, they seem to help us make sense of those experiences, to re-examine what we are about, and what we should do next. They seem to have some validity.

Acknowledgments

The research and development project reported in this paper was supported in part by the National Science Foundation under grant nos. MDR 897-0400 and MDR 885-0560. The development of the instrument to assess conceptual development and computational proficiency in arithmetic was supported by the Indiana State Department of Education. All opinions expressed are, of course, solely those of the authors.

References

Asch, S. E. (1952). *Social psychology*. Englewood Cliffs, NJ: Prentice-Hall.

Ausubel, D. P., Novak, J. D., & Hanesian, H. (1978). *Educational psychology: A cognitive view* (2nd

Ed.). New York: Holt, Rinehart, and Winston.

Barnes, B. (1977). *Interests and the growth of knowledge*. London: Routledge & Kegan Paul.

Carpenter, T. P., & Moser, J. M. (1984). The acquisition of addition and subtraction concepts in grades one through three. *Journal for Research in Mathematics Education, 15*, 179-202.

Cherryholmes, C. H. (1988). Construct validity and the discourses of research. *American Journal of Education, 96*, 421-457.

Cobb, P. (1986). Context, goals, beliefs, and learning mathematics. *For the Learning of Mathematics, 6*(2), 2-9.

Cobb, P. (1988). The tension between theories of learning and theories of instruction in mathematics education. *Educational Psychologist, 23*, 87-104.

Cobb, P., & Merkel, G. (1989). Thinking strategies as an example of teaching arithmetic through problem solving. In P. Trafton (Ed.), *New directions for elementary school mathematics: 1989 yearbook of the National Council of Teachers of Mathematics* (pp. 70-81). Reston, VA: NCTM.

Cobb, P., Wood, T., & Yackel, E. (in press a). A constructivist approach to second grade mathematics. In E. von Glasersfeld (Ed.), *Constructivism in mathematics education*. Holland: Reidel.

Cobb, P., Wood, T., & Yackel, E. (in press b). Analogies from the philosophy and sociology of science for analyzing classroom life. *International Journal of Science Education*.

Cobb, P., Yackel, E., & Wood, T. (1989). Young children's emotional acts while doing mathematical problem solving. In D. B. McLeod & V. M. Adams (Eds.), *Affect and mathematical problem solving: A new perspective* (pp. 117-148). New York: Springer-Verlag.

Cobb, P., Wood, T., Yackel, E., Nicholls, J. G., Wheatley, G., Trigatti, B., & Perlwitz, M. (1989). Assessment of a problem-centered second grade mathematics project. Unpublished manuscript. West Lafayette, IN: Purdue University, Department of Education.

Confrey, J. (1986). A critique of teacher effectiveness research in mathematics education. *Journal for Research in Mathematics Education, 17*, 347-360.

Covington, M. V. (1984). The motive for self-worth. In R. Ames & C. Ames (Eds.), *Research on motivation in education, Vol. 1. Student motivation* (pp. 77-113). New York: Academic Press.

Covington, M. V., & Beery, R. (1976). *Self-worth and school learning*. New York: Holt, Rinehart, and Winston.

Crutchfield, R. S. (1962). Conformity and creative thinking. In H. E. Gruber, G. Terrell, & M. Wertheimer (Eds.), *Contemporary approaches to creative thinking* (pp. 120-140). New York: Prentice-Hall.

Damon, W. (1981). Exploring children's social cognition on two fronts. In J. H. Flavell & L. Ross (Eds.), *Social cognitive development: Frontiers and possible futures* (pp. 154-175). New York: Cambridge University Press.

Dennett, D. C. (1978). *Brainstorms: Philosophical essays on mind and psychology*. Montgomery, VT: Bradford.

DeVries, R., & Kohlberg, L. (1987). *Programs of early education: The constructivist view*. New York: Longman.

Dewey, J. (1966). *Democracy and education*. New York: The Free Press.

Doise, W., & Mugny, G. (1984). *The social development of the intellect* (A. St. James-Emler, N. Emler, & D. Mackie, Trans.). New York: Pergamon (original work published in 1981).

Einstein, A. (1956). *Out of my later years*. New York: Philosophical Library.

Frederiksen, N. (1984). The real test bias: Influences of testing on teaching and learning. *American Psychologist, 39*, 193-202.

Frieze, I. H., Francis, W. D., & Hanusa, B. H. (1983). Defining success in classroom settings. In J. M. Levine & M. C. Wang (Eds.), *Teacher and student perceptions: Implications for learning* (pp. 3-28). Hillsdale, NJ: Lawrence Erlbaum.

Furth, H. G. (1981). *Piaget and knowledge: Theoretical foundations* (2nd ed.). Chicago: University of Chicago Press.

Helmke, A., Schneider, W., & Weinert, F. E. (1986). *Main results of the German IEA classroom environment study and implications for teaching*. Paper presented at the meeting of the American Educational Research Association, San Francisco.

James, W. (1907). *Pragmatism: A new name for some old ways of thinking*. New York: Longmans, Green, & Company.

Johnson, D. W., Johnson, R., & Scott, L. (1978). The effects of cooperative and individualized instruction on student attitudes and achievement. *Journal of Social Psychology, 104*, 207-216.

Johnston, P. (1989). Constructive evaluation and the improvement of teaching and learning. *Teachers College Record, 90*, 509-528.

Kamii, C. (1986). Place value: An explanation of its difficulty and educational implications for the primary grades. *Journal of Research in Early Childhood Education, 1*, 75-86.

Kelley, H. H. (1973). The process of causal attribution. *American Psychologist, 28*, 107-128.

Kohlberg, L., & Mayer, R. (1972). Development as the aim of education. *Harvard Educational Review, 42*, 449-496.

Kouba, V. L. (1989). Children's solution strategies for equivalent set multiplication and division word problems. *Journal for Research in Mathematics Education, 20*, 147-158.

Kuhn, T. S. (1970). *The structure of scientific revolutions* (2nd edition). Chicago: University of Chicago Press.

Lasch, C. (1978). *The culture of narcissism: American life in an age of diminishing expectations*. New York: Norton.

Maehr, M. L., & Braskamp, L. A. (1986). *The motivation factor: A theory of personal investment*. Lexington, MA: Lexington.

Maehr, M. L., & Nicholls, J. G. (1980). Culture and achievement motivation: A second look. In N. Warren (Ed.), *Studies in cross-cultural psychology*. New York: Academic Press.

McArthur, L. Z., & Baron, R. M. (1983). Toward an ecological theory of social perception. *Psychological Review, 90*, 215-238.

National Council of Teachers of Mathematics (1989). *Curriculum and evaluation standards for school mathematics*. Reston, VA: National Council of Teachers of Mathematics.

Nicholls, J. G. (1989). *The competitive ethos and democratic education*. Cambridge, MA: Harvard University Press.

Nicholls, J. G., Cheung, P. C., Lauer, J., & Patashnick, M. (1989). Individual differences in academic motivation: Perceived ability, goals, beliefs, and values. *Learning and Individual Differences, 1*, 63-84.

Nicholls, J. G., Cobb, P., Wood, T., Yackel, E., & Patashnick, M. (in press). Dimensions of success in mathematics: Individual and classroom differences. *Journal for Research in Mathematics Education*.

Nicholls, J. G., Licht, B. G., & Pearl, R. A. (1982). Some dangers of using personality questionnaires to study personality. *Psychological Bulletin, 92*, 572-580.

Nicholls, J. G., Patashnick, M., & Nolen, S. B. (1985). Adolescents' theories of education. *Journal of Educational Psychology, 77*, 683-692.

Nolen, S. B. (1988). Reasons for studying: Motivational orientations and study strategies. *Cognition and Instruction, 5*, 269-287.

Rorty, R. (1983). Method and morality. In N. Haan, R. N. Bellah, P. Rabinow, & W. M. Sullivan (Eds.), *Social science as moral enquiry* (pp. 155-176). New York: Columbia University Press.

Rosenholtz, S. J., & Simpson, C. (1984). The formation of ability conceptions: Developmental trend or social construction? *Review of Educational Research, 54*, 31-63.

Sharan, S., Kussell, P., Hertz-Lazarowitz, R., Bejarano, Y., Raviv, S., & Sharan, Y. (1984). *Cooperative learning in the classroom: Research in desegregated schools*. Hillsdale, NJ: Lawrence Erlbaum.

Sigel, I. E. (1981). Social experience in the development of representational thought: Distancing theory. In I. E. Sigel, D. M. Brodzinsky, & R. M. Golinkoff (Eds.), *New directions in Piagetian theory and practice* (pp. 203-218). Hillsdale, NJ: Lawrence Erlbaum.

Steffe, L. P. (1989, April). *Operations that generate quantity*. Paper presented at the annual meeting of the American Educational Research Association, San Francisco.

Steffe, L. P., & Cobb, P. (1984). Children's construction of multiplicative and divisional concepts. *Focus on Learning Problems in Mathematics, 6*, (1 & 2), 11-29.

Steffe, L. P., Cobb, P., & von Glasersfeld, E. (1988). *Construction of arithmetical meanings and strategies*. New York: Springer-Verlag.

Stigler, J. W., Smith, S., & Mao, L. (1985). The self-perception of competence by Chinese children. *Child Development, 56*, 1259-1270.

Taylor, S. E., & Brown, J. D. (1988). Illusion and well-being: A social psychological perspective on mental health. *Psychological Bulletin, 103*, 193-210.

Thorkildsen, T. A. (1988). Theories of education among academically precocious adolescents. *Contemporary Educational Psychology, 13*, 323-330.

Thorkildsen, T. A. (1989). Justice in the classroom: The student's view. *Child Development, 60*, 323-334.

Toulmin, S. (1983). The construal of reality: Criticism in modern and post modern science. In W. J. T. Mitchell (Ed.), *The politics of interpretation* (pp. 99-117). Chicago: University of Chicago Press.

Voigt, J. (in press). Social functions of routines and consequence for subject matter learning. *International Journal of Educational Research*.

von Glasersfeld, E. (1983). Learning as a constructive activity. In J. C. Bergeron & N. Herscovics (Eds.), *Proceedings of the fifth annual meeting of PME-NA* (pp. 41-69). Montreal: Psychology of Mathematics Education - North America.

Weiner, B. (1979). A theory of motivation for some classroom experiences. *Journal of Educational Psychology, 71*, 3-25.

Wood, T., & Yackel, E. (in press). The development of collaborative dialogue within small group interactions. In L. P. Steffe & T. Wood (Eds.), *Transforming early childhood mathematics education*. Hillsdale, NJ: Lawrence Erlbaum.

Yackel, E., Cobb, P., & Wood, T. (in press). Small group interactions as a source of learning opportunities in second grade mathematics. *Journal for Research in Mathematics Education Monograph*.

10

The Assessment of Schema Knowledge for Arithmetic Story Problems: A Cognitive Science Perspective

SANDRA P. MARSHALL

This chapter describes a cognitive science approach to the assessment of higher order thinking. The basic premise here is that to have meaningful changes in our tests, we need to bring into accord three central elements of the instruction and testing cycle: (i) our conceptions about the subject-area domain, (ii) our model of the student's learning in the domain, and (iii) our expectations about the ways that test items reflect the domain and its learning. In the following sections, a psychological theory of memory, learning, and instruction is outlined and applied to the domain of arithmetic story problems. Its implications for a corresponding model of assessment are discussed, and several examples of assessment items are given.

The Existing Problem

Suppose we have a highly simplified representation of the testing process (as in Figure 10.1). The central components are the subject matter to be tested, the student's knowledge about that subject matter, and the test that is intended to tap the student's knowledge. Many considerations come into play in this process. All are governed by our under-

standing of how the student learns, what he or she learns, and his or her ability to convey this learning. The test represents our best attempt to measure all three of these aspects. The test is the vehicle through which the student demonstrates his understanding.

There are clearly strong connections between the components of Figure 10.1. The test of knowledge, for example, should adequately sample the domain. However, an essential piece is missing from this figure; namely, an integrative model of memory. This model is needed to pull together the student's learning, the subject matter learned, and the test of the subject matter. It can be used to explain how and what knowledge the student encodes in the learning of the domain. It can be used to model the structure of the domain itself in terms of what we would like the student to know. And, it can provide the rationale for the test that estimates the student's learning about the domain. Thus, the model of memory serves as a framework on which the components of the testing process are developed. As shown in Figure 10.2, the model of memory drives the entire testing process by coordinating the different components around a common representation of knowledge. Its advantage lies in the way this representation

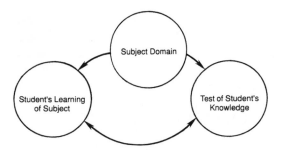

Figure 10.1. The testing process.

can be used. It can be used to guide our analysis of the domain, can provide the framework for our model of the student, and it can give us the basis for interpreting test performance.

At present, we do not generally employ an integrated model of the testing process as reflected in Figure 10.2. Rather, there are separate models for the domain and for the test, the model of a domain as a cohesive body of information and a testing model that depends upon the sampling of distinct, independent pieces of information. As a consequence, tests do not reflect the view of the domains that we wish to assess. This conflict can be described in terms of fragmentation and integration. We have confused integration of the domain with the aggregation or sum of many fragments of knowledge. That is, we have adopted a model in which we infer from the presence of many independent pieces of information that they are all well-connected and integrated within a student's mind.

The importance of this problem can be simply described in terms of graph theory.

Consider a graph in which there are several nodes or points (see Figure 10.3). These nodes may have few links, resulting in many fragments (Figure 10.3a), or they may all be linked together (Figure 10.3b). In a well-connected graph, any node can be reached ultimately from any other. In a graph with few connections, the paths are short and most nodes cannot be reached from a single starting point. Now suppose that we sample individually some nodes from a graph in which we have no information about connectivity. That is, we test for the presence or absence of each particular node. While the absence of a node does indeed indicate that it cannot be connected to other nodes, the presence does not conversely provide evidence that it is connected. In fact, we have no information at all about connectivity from such sampling. We know only that the collection of nodes is present in the graph. Full description of the graph demands at least two considerations: the number of nodes and the degree of connectivity among them. Our current testing reflects only the first of these, the number of distinct elements.

It is useful to represent both the domain and the student's knowledge of the domain as graphs. In our domain graph, we join together the important facts, concepts, strategies, procedures, and principles that govern the domain. A graph is particularly appropriate as a domain representation because it allows us to have a flexible model. Any particular node of information may be linked to many others, and there is no particular order in which several nodes are retrieved. The graph allows us to develop a network of information that can be accessed from any point.

A graph is also appropriate as our model for a student's knowledge of the domain. Its structure is consistent with recent neuropsychological research as well as research in cognitive science (e.g., semantic networks, parallel distributed processing, connectionism). The graph or network of student knowledge will generally be sparser than our ideal graph of the domain. As the student develops expertise in the domain, his graph will approach the ideal and will become more highly developed with a greater number of nodes and larger numbers of connections among them.

This graph representation allows us to

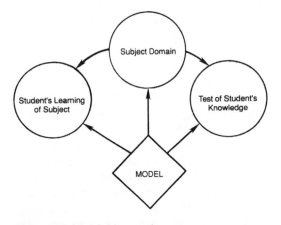

Figure 10.2. Model-driven testing process.

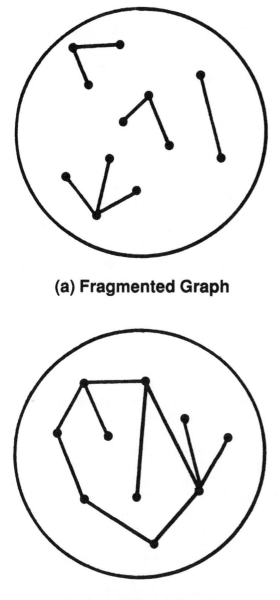

(a) Fragmented Graph

(b) Connected Graph

Figure 10.3 Simple graphs.

tions in student knowledge. To carry out this type of testing, we need a well-developed psychological model of learning and memory to capture what we mean by the "student graph" and to make useful estimations about what a student knows or doesn't know. The remainder of this chapter describes a particular model that has proven useful in this attempt.

The Psychological Theory

The models described in this chapter derive from schema theories of cognitive psychology (see, for example, Marshall, 1988; Rumelhart, 1980). Under such theories, human memory consists of networks of related pieces of information. Each network is a schema, a collection of well-connected facts, features, algorithms, skills, or strategies.

Cognitive scientists and psychologists have used the framework of schemas for positing how human memory is organized and for describing the mechanisms of learning.[1] Schemas are problem-solving vehicles. They aid in making sense of the environment, particularly of problems in the environment that have to be solved. They provide access to similar problems that have been encountered in the past, access to strategies that are relevant to diverse classes of problems, the means for reformulating or simplifying a problem, and the mechanisms for determining (and ordering) various goals and subgoals. In many problem-solving situations, the individual's choice of schema to govern the problem solving is at least as important as the particular solution obtained. Hence, the evaluation (or testing) of an individual's use of schema knowledge is necessary.

Schema theory

My own research has yielded an explicit formulation of the nature of a schema (cf., Marshall, 1989; in press). There are four primary components, each representing a different type of knowledge essential to schema development and usage. These four components are (i) feature recognition know-

consider a new perspective of testing: the degree to which the student has developed a well-connected body of domain knowledge. Clearly, if the domain is specified by nodes *and* connections among them, then tests of the domain should evaluate both the presence or absence of nodes and their existing connec-

---

1 Various names have been give to these formulations. For example, Minsky (1975) talks about frames, and Schank and Abelson (1977) describe scripts.

ledge, (ii) constraint knowledge, (iii) planning knowledge, and (iv) implementation knowledge. They are briefly described below; a more complete description may be found in Marshall (in press).

Feature recognition. This component carries the definitive characteristics, features, and facts that help one recognize the schema. Typically, one expects to have a prototypic example as well as a generic description of a situation to which the schema would apply.

Constraints. Within this component are rules and limitations having to do with the use of the schema. These constraints specify the necessary information that must be present for the schema to be invoked. It is convenient to think of this component as having slots into which parts of the current situation will be mapped.

Planning. Knowledge in this component concerns goals setting, building of expectations, and formulating a sequence of steps for solving the problem. Plans typically are top down in nature. Solutions, on the other hand, are bottom up.

Implementation. This knowledge is the set of procedures, rules, or algorithms that can be executed to reach subgoals and final goals.

Each of these components can be considered as a distinct network of interconnected pieces of knowledge. The more tightly connected the network, the easier the access to any part of it. The components themselves are also connected in a larger network; in a well-developed schema, any component will be reachable from any other. By the same argument, the schemas are linked together within the entire subject area in the memory of a good problem solver.

As a consequence of this structure, there are several levels and many parts to consider in schema assessment. We can ask questions about a single component, such as whether the feature-recognition component is fully developed. We can query the connections within a single schema by asking whether two or more specific components are linked. Or, we can investigate the cohesiveness of many schema networks by asking whether schemas are linked.

Comparison with other rationales

How does the cognitive rationale described here differ from rationales underlying existing tests? The principal difference is its focus on the integration of knowledge within human memory and its emphasis on testing degrees of connectivity. Most items on existing standardized instruments test only isolated pieces of information — usually single elements of the feature-recognition component or single algorithms of the implementation component.

Sampling issues. Test development is normally driven by sampling theory. It assumes that the test items are drawn as a random sample from the domain of knowledge to be tested. For example, one might wish to evaluate a child's prowess in adding pairs of single-digit numbers. Given the values 0–9, there are 45 different items one could choose (or 90 items if order within an item is important). Obviously, an estimate about the child's performance based upon one item would be suspect. Equally obvious is that one would not wish to have a child respond to all 45 (or 90) items. The solution is to make an estimate based upon some sampling of the items. One can even determine how many items are necessary in order to make valid estimates. Any sample of items is expected to serve equally well, for underlying this sampling rationale is the assumption that all items are equally representative of the domain.

Now consider a schema-based model of testing. Different items test connectivity at different levels (e.g., within a component network, within a schema network, or within the domain network). All levels are important in assessing schema knowledge, but the items testing them will not be parallel. They cannot be regarded as samples drawn at random from a uniform population. This is a fundamental difference in testing a domain under the two rationales. Under the sampling model, the elements within the domain are of equal importance. Under the schema theory, they are not.

Independence and priming. Other assumptions in standard test theory are important psychologically. It is usual to assume that items are independent. This means that the ability to respond to one item does not depend upon the ability to respond to another. Assume that we have only two items, each testing a different piece of knowledge. If these pieces of knowledge are independent, a student's memory retrieval of the first (and corresponding response to the test item about it) does not depend upon the presence of the second. If, on the other hand, the pieces of information are linked in a schema, the access and retrieval of one of them necessarily results in access to the other through a process called activation. This activation has a bearing on the dependence or independence of the items.

Activation is an important feature of a network. Whenever any constituent of the network is accessed, all other parts of the network that connect to it are to some degree activated. Consider the nature of activation. The human memory contains a very large number of bits and pieces of knowledge. They are in storage and are not easily accessed (e.g., we all "know" many facts that somehow we cannot retrieve). Our attempts to retrieve specific facts result in activation of those pieces of knowledge that are found in the retrieval process. Those that are activated are considered to be part of working memory, and only the knowledge contained in working memory appears to be accessible for immediate cognitive processing.

Now, assume that the fact you retrieve is completely isolated and unrelated to any other. Activation of it does not result in activation of any other knowledge. Only the single element will move into working memory. In contrast, assume that the fact is part of a rich network of knowledge that contains names, dates, relationships, and concepts. Activation of the target fact will activate the connected elements and will result in the placing of some (or all) of them into working memory. Now, they too can be used immediately without researching memory. This, of course, is a highly simplified version of activation, designed to point out the importance of connections. For an excellent discussion of how activation takes place, see Anderson (1983).

Relevant to the discussion here is a particular aspect of activation called priming. Priming has been extensively studied in the psychological literature. It refers to the improvement of retrieval of a piece of information when a related piece has been immediately activated. Thus, to solve the first of our two items, one piece of knowledge has been accessed and retrieved (for direct solution), and the second piece has been activated because it is associated with the first. This second piece is now primed and ready for additional cognitive processing when the second item is presented. The individual having the connected network in this case will be more likely to recognize and retrieve the correct response to the second item. The individual having fragmentary knowledge will need to search memory again for his response. The student who has only one of the two pieces of information is at a disadvantage as he or she tries to solves the second of the two items.

Declarative and procedural knowledge. Elementary mathematics tests tend to have two basic types of items: items that require a computation and items that require identification. Examples of the first are expressions such as $3 + 6 = \underline{\hspace{1cm}}$ and $234 \times 1.35 = \underline{\hspace{1cm}}$. Examples of the second are requests to select the name of a geometric shape or to identify features such as parallel lines. The computation items assess the student's ability to execute a set of rules or procedures. The identification items assess recognition of particular facts, concepts, or characteristics.

These two types of items are related to two well-known psychological concepts of memory: procedural knowledge and declarative knowledge. Procedural knowledge consists of if-then statements that can be processed by evaluating the "if" portions. Whenever the "if" is true, the "then" is carried out. Arithmetic algorithms of addition, subtraction, multiplication, and division are presumed to be encoded in memory in this fashion. On the other hand, some mathematics knowledge requires no computation and needs only to be looked up. This is declarative knowledge. Either the student has previously encoded the required knowledge into memory or he or she has not. The naming of shapes such as circle or triangle is one example.

Schema theory encompasses both declarative and procedural knowledge. Much of feature recognition is stored declaratively. Similarly, constraints and implementation knowledge concern procedures that have been learned. Schema-based assessment goes beyond the assessment of declarative and procedural knowledge to include evaluation of how declarative knowledge and procedural knowledge are used simultaneously in problem solving.

Implications for testing

What implications does the cognitive science rationale have for mathematics testing? There seem to be two important ones. First, we need to move away from the domain-sampling theory. Why should we want to make a random sample of a child's knowledge? Such a goal implies that we wish children to have broad, superficial knowledge about a large number of possibly unrelated topics. Is this the goal of our instruction? I hope not. Moreover, this rationale is not consistent with our view of an integrated body of knowledge within the domain.

A second implication that follows from the cognitive science perspective is that there may be several correct answers to an item, depending upon the knowledge structures that the student has thus far developed. For example, if the student is asked to describe a particular concept, he or she might respond by giving an example used in class, by giving an example created by the student, or by giving an abstract definition. All three are correct, but they indicate differing levels of sophistication. The point of this discussion is that the traditional view of "one right answer" for any test item is insufficient. Students have differing levels of understanding. Partial or limited understanding should not be penalized. Not all test items should be limited to the format of a single correct response.

Why have educators and psychologists developed tests having the traditional structure? Primarily, it has been done for statistical reasons. If one has a fixed number of test items that are random, equally important, and independent, one can weight them equally, score each of them as correct or incorrect, and aggregate the scores. One can then compare these totals, compute means and standard deviations, and carry out statistical tests. However, these procedures are limited indicators of the understanding of students who responded to the items.

An Example: SPS

In this section, I describe the use of schema theory as a basis for a computer-assisted instructional system. The system is SPS, the Story Problem Solver, an instructional system that I and my associates have developed over the past three years under support from the Cognitive Sciences Program of the Office of Naval Research.[2] SPS provides instruction to adult students about understanding arithmetic story problems. It has been used with community college students enrolled in remedial algebra classes and with beginning college students whose mathematics background is weak.

Although SPS was created to test a specific psychological theory of memory architecture, it has proved beneficial in developing new ideas about evaluation. When we produced a new type of instruction, it was necessary that we also develop a new type of test to evaluate the impact of that instruction. In particular, we constructed a collection of items to assess the development of schema knowledge for arithmetic story problems. These items focus upon the four components of knowledge described earlier. They follow instruction that is designed to promote the development of rich networks for each component.

SPS instruction

The objective in SPS instruction is to present information in such a way that it facilitates the student's creation of strongly connected schemas about arithmetic story problems. The underlying foci for the schemas are five semantic relations describing the relationships that may exist in story problems. The domain is depicted as consisting of five primary problem-solving schemas based upon the semantic relations. The instruction centers on the semantic relations and highlights important feature recognition, constraints, planning knowledge, and implementation strategies relevant to each one. It is expected that students completing the entire instructional sequence will have developed schemas based upon the semantic relations and will be able to use these schemas to solve complicated story problems.

The semantic relations of SPS are

---

2 Additional details are given in Marshall, Barthuli, Brewer, & Rose, 1989.

Change, Group, Compare, Restate, and Vary. Students receive instruction about each of these. They work with the relations by name and have a number of different instructional sessions in which they focus upon important features of the relations, in which they learn about the conditions necessary for each to occur, and in which they develop the ability to nest the relations within multi-step story problems. An important component of SPS instruction is a set of diagrams that are used to illustrate the relations and the necessary parts within each one.

Schema assessment in SPS

Five types of items are used to assess schema knowledge in SPS. Each is summarized here. The reader should disregard the particular content of the items, which depends upon the instruction presented in SPS. The items are presented here as examples of various formats that can be used to assess schema knowledge. Additional details about the items and the instruction are given elsewhere (Marshall et al., 1989).

The first item type is a recognition task that assesses the complexity of feature knowledge. The format is not new, although it is rarely found on traditional mathematics tests. Figure 10.4a contains a recognition item. The objective is for the student to recognize the situation that is described in the problem. The

student is not asked to compute a numerical solution. By shifting the emphasis away from computation, SPS helps the student focus upon the importance of understanding the relations expressed in the problem and how these relations are embedded in the story situation. An essential part of that understanding is the recognition of key features.

Figure 10.4b is an alternative version of the first item type. Here, the student responds by selecting the diagram representing the situation instead of selecting the situation name.

A second item type stresses understanding of the different elements that may be found within a single schema. For this task, SPS makes use of the diagrams developed and used during the computer-based instruction. An important objective in SPS is for the student to develop appropriate schemas of the basic situations that appear in story problems. For each schema and its corresponding situation, there are specific constraints that apply. Central among these constraints are the distinct elements or parts that make up the situation. The diagrams emphasize these parts. A student eventually should develop the competence to map a new problem into the appropriate schema without the aid of an external diagram. During the initial stages of instruction, the diagrams serve as the means of making explicit the relationships expressed in

EXPLANATION	ITEM
Instructions to the student	**INSTRUCTIONS:** Read the story below. Decide which of the five situations best describes the story. When you have made your choice, position the arrow on the top of the one you have selected and click the mouse button once.
Problem to be solved	Gina collects stuffed animals. She has 11 stuffed animals in her room. Of these there are 3 rabbits and 2 frogs, and the rest are bears. This means that she has 6 bears in her stuffed animal collection.
Student response Note: Student responds by selecting an option and using a mouse to indicate the selection. The correct response is "Group."	Change Group Compare Vary Restate
	OKAY REVIEW

Figure 10.4a. Assessment item for feature recognition knowledge: Name.

EXPLANATION	ITEM

Instructions to the student

INSTRUCTIONS: Choose the one diagram below that fits this story problem. Move the arrow into the diagram you have selected and click the mouse button.

Problem to be solved

Right now in the grocery store, Red Delicious apples cost 89 cents per pound. Jonathan apples are 40 cents a pound less than the Red Delicious apples, so they must cost 49 cents per pound.

Student responds by selecting one of the diagrams and using a mouse to place them in the diagram.

Correct response is the lower left figure which represents "Restate."

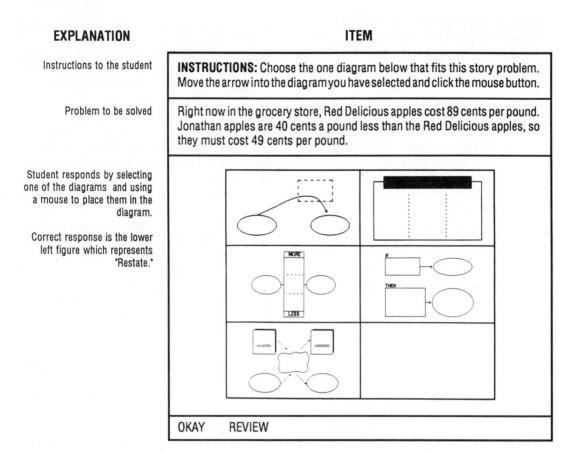

OKAY   REVIEW

Figure 10.4b. Assessment item for feature recognition knowledge: Diagram.

the problems. Consequently, the test items for the constraint knowledge consist of problems presented with the correct diagrams. Students are asked to move pieces (e.g., words or phrases) from the problems into the appropriate parts of the diagrams. Figure 10.5 contains an example of such items.

A primary concern of the cognitive science perspective is the integration of knowledge. In the realm of arithmetic story problems, this integration is reflected in multistep problems. A large part of SPS's instruction focuses on multistep problems, emphasizing the need to recognize the predominant situation (the problem) of the story and to identify any secondary situations (the subproblems). Equally important is the recognition of how the schemas—and consequently the diagrams—fit together. Understanding of multistep items is assessed in two ways. First, students are simply asked to recognize and name the situations that correspond to the overall (or primary) question

posed in the problem and to the embedded (or secondary) question that must be answered before finding the overall answer. This task draws upon the student's ability to formulate a plan. The student must determine through feature knowledge and constraint knowledge whether a particular schema is appropriate. If it is, the student then must formulate a plan for reaching a solution. Since these are multistep items, the plan will call for the solution of an embedded subproblem prior to the solution of the overall problem. As an initial attempt to assess students' understanding of the overall and secondary situations, SPS uses problems requiring two steps for solution and has the student identify the appropriate situation for each step (see Figure 10.6). Again, numerical solution is not involved.

The other item type focusing on multistep problems also requires planning by the student. This task involves the diagrams. The student is asked to identify which parts of the diagram can be filled by information that is

EXPLANATION	ITEM

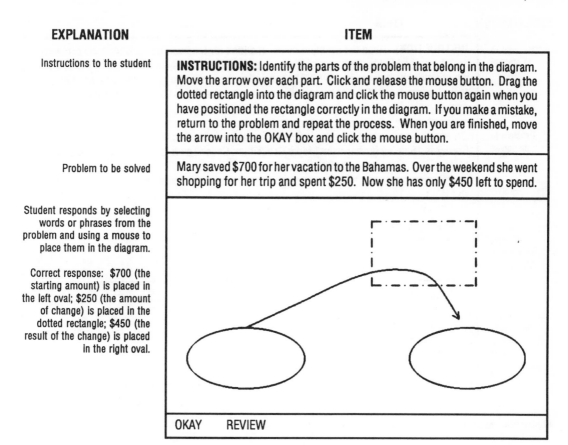

Instructions to the student

INSTRUCTIONS: Identify the parts of the problem that belong in the diagram. Move the arrow over each part. Click and release the mouse button. Drag the dotted rectangle into the diagram and click the mouse button again when you have positioned the rectangle correctly in the diagram. If you make a mistake, return to the problem and repeat the process. When you are finished, move the arrow into the OKAY box and click the mouse button.

Problem to be solved

Mary saved $700 for her vacation to the Bahamas. Over the weekend she went shopping for her trip and spent $250. Now she has only $450 left to spend.

Student responds by selecting words or phrases from the problem and using a mouse to place them in the diagram.

Correct response: $700 (the starting amount) is placed in the left oval; $250 (the amount of change) is placed in the dotted rectangle; $450 (the result of the change) is placed in the right oval.

OKAY   REVIEW

Figure 10.5. Assessment item for constraint knowledge.

already known from the problem, which part(s) can be known only by solving embedded or secondary problem(s), and which part corresponds to the overall question posed in the problem. Figure 10.7 shows such an item.

Finally, the fifth item type used to assess student knowledge in SPS focuses on the student's ability to select the appropriate arithmetic operation, given the particular situation. These items follow instruction about the different situations and the various questions that can be framed from any one of them. A central part of this instruction tackles the issue of key word reliance. We expect the students to use their understanding of key words and yet to avoid the pitfalls of relying exclusively upon them. They can do this by keeping their attention on the nature of the situation. The items used in this assessment are simple one-step items, and students select the appropriate mathematical formulation, equation, or expression that leads to a correct

solution. Figure 10.8 contains an example.

These five item types provide very different information about a student's problem solving than is gathered from traditional items, which seek only a numerical solution. From the first type, we gain insight into the student's development of feature knowledge for each schema. From the second, we determine whether the student understands adequately the different parts of the situation and how they fit together. This task requires constraint knowledge. From the third and fourth, we see whether the student can formulate plans and link together several schemas. Finally, the last item type tests the student's implementation knowledge.

One can contrast these items with a typical test item: a multistep story problem that requires the student to select a numerical solution from a set of numerical options. What information does such an item provide about the student's understanding? Consider a correct response. Suppose the student made the

EXPLANATION   **ITEM**

Explanation	Item
Instructions to the student	**INSTRUCTIONS:** Read the problem below and look for the overall situation in the problem and the embedded secondary situation. First, move the arrow to the name of the overall situation (in the box marked OVERALL) and click the mouse button. Next, move the arrow to the box marked SECONDARY and click on the situation that is the secondary or embedded one.
Problem to be solved	Greensville is 40 miles from Maple Grove. Cedar Town is 13 miles from Maple Grove. Oak Corner is 15 miles farther from Maple Grove than Cedar Town is. Which town is closer to Maple Grove, Greensville or Oak Corner?

First response by student: Select overall situation. Correct response is "Compare."	OVERALL
	    Change
	    Group
	    Compare
	    Vary
	    Restate

Second response by student: Select secondary situation. Correct response is "Restate."	SECONDARY
	    Change
	    Group
	    Compare
	    Vary
	    Restate

Figure 10.6. Assessment item for planning knowledge (general structure).

correct selection. We infer that he or she reasoned correctly, although we have no supporting evidence. We qualify our inference to take into account the possibility of guessing. What if the student was incorrect? Our inferences now depend critically upon the possible distractors. Rarely do the distractors offer choices that reflect upon the different types of knowledge needed to use a schema. More typically, they are superficial errors of computation. The student who has difficulty with feature recognition, for example, will not necessarily encounter a distractor that seems reasonable to him. His response may be to guess or to skip the problem altogether and go on to another one.

If we want to have information about how students are processing information and about their ability to integrate and use their knowledge, we need to construct test items that will give us specific insights into these abilities. One alternative is to construct items, such as those described earlier, which ask questions about intermediate steps in problem solving. Another option is to ask general ques-

tions and interpret the responses in terms of a well-developed theory of learning. An example of this is described in the following section.

Related assessment of SPS and schema development

The items described previously are implemented as part of the instruction and assessment of student learning in a computer environment. As a supplement to that evaluation, students using SPS have also responded to paper-and-pencil questions and to oral questions posed during follow-up interviews. Of particular importance here are a set of broad questions that allow the student to demonstrate his or her understanding of the general structure of the semantic relations. The questions are broad and open-ended (e.g., "What can you tell me about Change?"), and there are many acceptable answers. After a single computer-instruction lesson, virtually all students can give a reasonable response to the questions. However, their depth of understanding varies. This variation is easily seen in

EXPLANATION   **ITEM**

Instructions to the student

INSTRUCTIONS: Read the problem below and study the diagram. For each part of the diagram, decide whether the necessary information is already GIVEN in the problem, whether you can find it by first getting a PARTIAL ANSWER, or whether you can find it as the FINAL ANSWER to the problem. Fill each part of the diagram with one of the three choices. Click in the OKAY box when you have filled the diagram.

Problem to be solved

Julie had a budget of $1200 to furnish her new apartment. She found a five-piece living room set on sale for $625. She also found a queen-sized bed for $350 and a dresser for $195. How much money, if any, will Julie have left to buy miscellaneous odds and ends for her apartment?

Student responds by selecting words or phrases from the problem and using a mouse to place them in the diagram. Correct response: *Left Oval:* GIVEN (starting amount, $1200). *Rectangle:* PARTIAL ANSWER (amount of change can be computed from information in problem). *Right Oval:* FINAL ANSWER (the overall unknown of the problem).

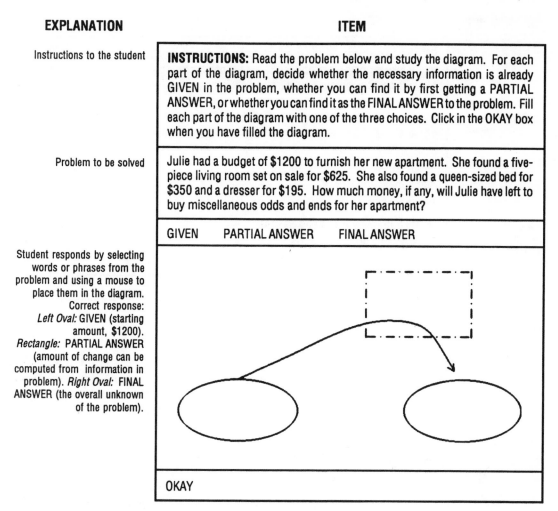

GIVEN   PARTIAL ANSWER   FINAL ANSWER

OKAY

Figure 10.7. Assessment item for planning knowledge (specific components).

their responses.

My associates and I have observed three general types of response to questions such as "What can you tell me about Change?" First, many students can recall only the initial example used in SPS to introduce the semantic relation. Some students describe Change by using one of the other examples developed later in SPS. Occasionally, a student will use one of the exercise problems in giving the description. A relatively rare response is the generation of a unique example to illustrate the relation. For all three of these responses, students are unable to give a generalized description of the relation and can only talk about the relation with respect to an example. Additional questions to them elicit no further information.

A second response observed in our stu-

dents is the formulation of an abstract statement about change, such as "A change occurs whenever you have an increase or a decrease of something." Students who give this response usually cannot embellish it beyond supplying an example.

Finally, we have also found that some students can give a precise description of a relation in terms of the conditions and constraints that define it. An example of such a response is "A change happens over time and involves three things: an initial amount, the amount of change, and the final amount." These students are able to give examples and to elaborate on each part of the description.

From the students' responses and from the schema theory used to build SPS, we can make the following inferences about students' understanding of the semantic relations. First,

EXPLANATION　　　　**ITEM**

Instructions to the student	**INSTRUCTIONS:** Read the problem below and then look at the possible steps you might take to solve the problem. Select the one that will solve the problem correctly. Move the arrow to your selection and click the mouse button.
Problem to be solved	Alice went to the grocery store with $35.00. She purchased her groceries and went home with $19.00. How much did she spend at the grocery store?
Student response	**POSSIBLE STEPS:** Add $35.00 and $19.00. Multiply $35.00 by 2 and add $19.00. Divide $35.00 by 2 and add $19.00. Subtract $19.00 from $35.00. OKAY

Figure 10.8. Assessment item for implementation knowledge.

the response of an example is the most primitive and indicates the least understanding. As they gain additional insight and familiarity with the relations, students move to a general description of the relations. Once they reach this level, they do not retreat back to the example stage. We have traced the development of understanding in several groups of students and have not observed a single instance of abstract description followed in later questions by example description without the abstract as well. The description in terms of constraints appears to indicate still greater understanding on the part of the student. With this response, the student is attempting to provide sufficient information to describe fully the situation in which a particular relation may occur. Again, we see no backtracking once students are able to provide this level of description.

In terms of schema development, it appears that students initially learn by first encoding in memory a salient example that illustrates it. This is often an example from instruction. As they gain understanding, students add to the knowledge network in memory and develop examples of their own, usually derived from their own experiences. These examples show the integration of the new information with existing information previously stored in memory. With yet more instruction, students are able to develop an abstract characterization. At this level of understanding, students recognize the similarities among several examples and can articulate them in a general statement.

I have followed students in the lab and can trace this path of schema development (Marshall, in press). Students' abilities to formulate their understanding correlate positively with their abilities to respond to the assessment items presented as computer exercises. Students who possess a greater understanding in terms of schema development perform better on the assessment items than do students with weaker understanding.

The importance of these questions from an assessment perspective is that there are several correct responses that can be made to them. Thus, the value is not in judging whether or not a response is correct but rather to evaluate the response in terms of the level of understanding demonstrated in it.

General Usefulness

Much of mathematics instruction and its subsequent assessment centers on computational skills. In arithmetic, considerable time is devoted to the learning and practice of the four operations. In algebra, similar attention is paid to factoring polynomials. In statistics, emphasis is upon computation of different techniques and statistical tests. The advent of calculators in the elementary and high schools and of computers in the university make these three emphases obsolete.

Consider the case of arithmetic. In the near future, every child will have access to a calculator. Whether the child can correctly execute the long-division algorithm will be less important than whether he or she recognizes the need to divide. More important will be the child's grasp of number, basic number facts, and the ability to evaluate whether a result is reasonable. Many tests are now being developed in which it is assumed that all children will have the option of using hand-held calculators. Unless the test items are truly trivial, calculators will make a difference in children's performance. Errors in carrying out the arithmetic algorithms will decrease (since calculators don't make arithmetic errors). We will need test items that tap into children's understanding of what the operations are, when they are appropriate, how (e.g., in what order) to combine them, and how to carry them out (e.g., by hand, by calculator, by mental calculation). Notice that these issues correspond to the four aspects of schema knowledge presented earlier. Knowing what multiplication means and something about its uses (cf. Usiskin & Bell, 1983) falls under feature-recognition knowledge. The issue of when it is appropriate to multiply comes under constraint knowledge (e.g., are there sufficient pieces, are they on a similar scale, is multiplication a legitimate operation given the elements of the problem?) When several operations are to be carried out, planning knowledge is needed. Finally, execution of some algorithm (either hand or machine calculation) is necessary to reach a numerical solution.

One can assess any of these issues in a manner similar to that described for story problems. To do so requires specification of the important schema knowledge of the domain of arithmetic operations and delineation of the specific knowledge contained in each of the four components. The critical requirement is the development of the model of the assessment process in conjunction with the model that characterizes both the domain and

the student's learning of the domain, as in Figure 10.2.

The approach outlined in this chapter is not limited to assessing knowledge of arithmetic. A parallel case could be made for algebra or statistics. No longer do students (or researchers) carry out statistical tests by hand. Computers are routinely used, frequently accessed by complex statistical packages. Users have greater need for understanding the assumptions and limitations of individual statistical techniques than for memorizing a large set of equations. Why should students be tested by asking them only to work out by hand the computations involved in a t-test, analysis of variance, or regression problem? Is it not more valuable to ascertain whether they understand the basic model that is being tested, the severity of violating certain assumptions, the interpretation of the statistical findings, and the extent to which their findings generalize? Schema-based assessment can offer tests of these abilities.

Conclusions

There are many calls today for new tests of mathematics and science. Psychological models of cognition and cognitive processing can provide test developers with the basis for constructing these new tests. Schema-based assessment already provides a wide variety of options to mathematics educators and test developers. The underlying schema theory is a means of organizing the domain of knowledge to be assessed. By its very nature, the schema changes the focus from individual pieces of knowledge to a coherent network.

Many educators have called for tests of "higher order thinking." What does such thinking encompass? Thus far, there are many broad descriptions, but few operational definitions. Models such as the schema theory described here are can make these ideas operational if educators take the lead in adopting them as a basis for instruction and if test developers incorporate them into tests.

An advantage of the schema theory described here is that it provides the means of incorporating cognitive science and mathematics assessment. It links readily with smaller units (as studied in neuroscience models) as

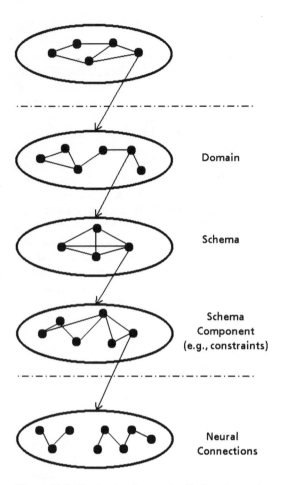

Domain

Schema

Schema
Component
(e.g., constraints)

Neural
Connections

Figure 10.9. The levels of networks. Each node can be expanded into a lower level network.

well as with broad units at the curriculum level. Figure 10.9 provides a general picture of these relationships. One can focus the assessment on different levels of knowledge, depending upon the goal of assessment, ranging from individual schemas to the integrated curriculum.

The schema theory presented here already has served as the basis for an instructional system and its accompanying assessment. It is a proven way to focus assessment upon the integration of knowledge rather than on fragmentary knowledge. It provides a basis for improving the assessment

of mathematical thinking and for incorporating cognitive science into evaluation in related fields. Its value is that it provides the coordination for representing what we want to test, how a student learns it, and how the test measures it.

Acknowledgments

The research described in this chapter was carried out under support from the Cognitive Sciences Program, Office of Naval Research, Contract No. N00014-K-85-0661.

References

Anderson, J. R. (1983). *The architecture of cognition.* Hillsdale, NJ: Erlbaum.

Marshall, S. P. (in press). *Assessing schema knowledge.* In N. Frederiksen (Ed.), Test theory for a new generation of tests. Hillsdale, NJ: Lawrence Erlbaum.

Marshall, S. P. (1989). Affect in schema knowledge: Source and impact. In D. B. McLeod & V. M. Adams (Eds.), *Affect and mathematical problem solving: A new perspective (pp. 49-59).* New York: Springer-Verlag.

Marshall, S. P., Barthuli, K. E., Brewer, M. A., & Rose, F. E. (1989). *Story Problem Solver: A Schema Based System of Instruction.* (Tech Rep. 89-01, Contract No. N00014-85-K-0661). Arlington, VA: Office of Naval Research.

Minsky, M. (1975). A framework for representing knowledge. In P. H. Winston (Ed.), *The psychology of computer vision.* New York: McGraw Hill.

Rumelhart, D. E. (1980). Schemata: The building blocks of cognition. In R. Spiro, D. Bruce, & W. Brewer (Eds.), *Theoretical issues in reading comprehension.* Hillsdale, NJ: Erlbaum.

Schank, R., & Abelson, R. (1977). *Scripts, goals, and understanding.* Hillsdale, NJ: Erlbaum.

Usiskin, Z., & Bell, M. (1983). *Applying arithmetic: A handbook of applications of arithmetic. Part II: Operations.* The University of Chicago, Department of Education, Arithmetic and Its Applications Project.

11

Critical Evaluation of Quantitative Arguments

CURTIS C. McKNIGHT

Higher order thinking, even in mathematics, is not a unitary phenomenon. There is not one form of higher order thinking but, rather, many forms. It might be argued that each form is the same complex of cognitive processes applied to differing domains of knowledge and tasks. However, as seen in this volume, the task and knowledge domains differ so greatly that, even if there are important commonalities in the cognitive processes involved in complex thinking in those domains, pragmatically, it may be more useful to consider the thinking tasks of each domain in terms of their specificity, with a focus on the tasks particular to the knowledge of the domain and the forms of higher order thinking critical to such tasks. Such is the case with the critical evaluation of quantitative arguments.

The entrance of our society into the information age, with its ubiquity of computer-enhanced publishing and presentation graphics, has led to a virtual bombardment of both citizen and student with numerical data and, moreover, with numerical data embedded not in the context of separate treatises or texts on mathematics, but in the context of informative articles which often contain either central or peripheral arguments that have essential quantitative elements. In particular, the quantum leap in the ease with which presentation graphics can be generated has resulted in constant exposure to information presented in graphical form. The ability to think critically in the presence of arguments with essential quantitative elements, often graphical elements, has become an essential skill for educated citizens in our society and will be so even more in the future. Instruction related to the skills necessary for such quantitatively oriented critical thinking will certainly enter the curricula of school mathematics if the new standards for mathematics as reasoning and communication promulgated by the National Council of Teachers of Mathematics (NCTM, 1989) have their hoped-for impact.

Even with the emergence of these new essential skills of critical thinking that utilize tools for quantitative reasoning and graphical interpretation in contexts that are not "mathematical" in the narrow or disciplinary sense, little investigation has yet been done of the nature of these cognitive skills, their characterization in information-processing terms, and their interaction with noncognitive factors such as anxiety about mathematics. Anecdotal evidence suggests that mathematics anxiety results in the paralysis of critical thinking abilities when quantitative elements are included in the contexts in which individuals must think critically. Further, it has been suggested that the presumed fact of this paralysis of critical facilities, along with the wide spread of such mathematics anxiety, provides at least the possibility for use of numerical data and presentation graphics precisely to eliminate critical evaluation of arguments and claims.

Standard texts on presentation graphics (e.g., Lefferts, 1981) list a variety of uses for such graphics: clarifying, simplifying, emphasizing, summarizing, reinforcing, arousing interest, providing impact, increasing credibility, and enhancing coherence (p. 19). To this catalog of legitimate uses of graphics and numerical data must, unfortunately, be added their perversion in arousing interest without providing appropriate substance, serving the cause of credibility when such credibility is not merited, producing impact sufficient to overcome innate critical responses, and so on.

Certainly, considerable attention has been given to the direct investigation of mathematical problem solving (e.g., Mayer, Larkin, & Kadane, 1976; Lester, 1982; Schoenfeld, 1985). Such studies, however, have often been in isolated contexts, or ghettos, of directly mathematical tasks. Far less investigation has been done of the mathematical tasks that arise in essentially "nonmathematical" contexts. An extensive research literature exists on problem solving more generally (e.g., Greeno, 1978; Larkin, McDermott, Simon, & Simon, 1980; Mayer, 1983; Polson & Jeffries, 1982; Rowe, 1985; Simon, 1978). Often, these studies have used mathematics as a content of convenience or have used problems from scientific contents (e.g., physics) that have essential mathematical aspects. Far less often, however, have such studies used contents from the social sciences and even less often from the sorts of popular media typically encountered in mass culture.

Yet another literature exists more generally on intelligence and on reasoning in general, reasoning conceptualized as the cognitive processes of literate individuals encountering the (typically nonnumerical) data and arguments to which educated individuals are exposed. For instance, a special issue of the *Review of Educational Research* was devoted to just these issues of teaching and learning reasoning skills (see Glasman, Koff, & Spiers, 1984). In this issue, reasoning and learning to reason were examined in a variety of contexts — for instance, learning through writing (Applebee, 1984) or reasoning in the context of a (nonmathematical) test (Haney, 1984). In other studies, the rise of scientific reasoning itself has been more directly examined (Siegler, 1978). Yet, in each of these more general contexts in which "reasoning," in contrast to "problem solving," is examined, situations with essentially quantitative elements are almost wholly ignored, even as a content of convenience for such studies.

The importance of the impact of situations with essential quantitative elements demanding a reasoned and critical response, together with the lacunae in the relevant research literatures, suggest that it is time — and past time — for mathematics education researchers, especially those with a concern for higher order thinking skills, to mount a direct attack on the questions of cognitive characterization of such thinking performances in the interests of providing the bases for improved teaching and learning of skills essential for educated citizens. The tools for such investigation are available.

The literature on analyzing cognitive processes (e.g., Franks, Bransford, & Auble, 1982; Kail & Bisanz, 1982) and for the use of human information-processing approaches in task analyses that characterize cognitive tasks (e.g., Gardner, 1976; Gregg, 1976; Resnick, 1976; Ericsson & Simon, 1984) are, in many cases, not only available but have been in use for over a decade. Even the more specialized literature on what may well be revealed as an essential set of cognitive skills, that of metacognitive awareness, has had its groundwork laid in foundational studies that are now over 10 years old (e.g., Brown, 1975; Flavell, 1976).

What remains, then, is to bring these appropriate literatures and tools together into a direct assault on pressing questions of interest that as yet have had little direct investigation. Such a foray would add to the agenda of mathematics education research the investigation of the critical evaluation of quantitative arguments. What this chapter seeks to provide is, in a sense, a feasibility study for adding these questions to this research agenda and for exploring some of the issues involved. The study seeks to be exploratory and illuminating, to sketch broadly rather than to be definitive and conclusive. The investigation of the questions considered here is likely to take the efforts of many researchers over a considerable period of time. What is attempted is simply a beginning adequate to show the importance of the questions and feasibility of direct research to answer those questions.

The Survey

The remainder of this chapter presents a consideration of the method and results of a brief survey of critical evaluation of graphical arguments (i.e., arguments that have an essential graphical element). What is involved is a series of situations in which data are presented in graphical form, which may or may not have useful value as evidence for the truth of a stated proposition related to that graphical data. Subjects are asked a series of questions to assess their ability both to interpret such graphical data and to decide their evidential value in demonstrating the truth of the stated proposition. The sample is sufficiently small and the instrument filled with questions that sketch sufficiently broadly that it must be reiterated that all that was intended by this survey was a type of feasibility study, to assess the value and guide the direction of later, more detailed research.

The instrument

The instrument consisted of four cases, each involving a single graph and a related proposition. A deliberate effort was made to select graphical presentations that would be encountered either in relatively popular media (e.g., *Scientific American*) or typical texts and monographs. A further effort was made to include graphs for which there were serious interpretational difficulties, as well as those which were more straightforward to interpret. Finally, some graphs were related to propositions that were patently false ("storks bring babies") and others that would seem more likely to be true ("populations will increase faster in developing countries than in developed countries"). By introducing these sources of variance, it was hoped to learn both something about the impact of interpretational questions on evaluating the evidential value of data and about individuals' abilities to separate the particulars of a given (likely false) proposition from more general criteria for what constitutes useful evidence for the truth of a proposition.

The first case utilized a graph that related the number of pairs of brooding storks to the number of newborn human babies in a single province of West Germany for a few years. This graph appeared in *Scientific American* as

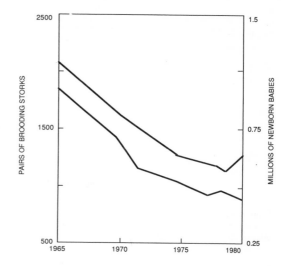

Figure 11.1. Numbers of storks versus numbers of newborn babies. (Simplified graph adapted from *Scientific American*.)

a humorous element in the "Science and the Citizen" column and a simplified adaptation is seen in Figure 11.1. Critical to the graph are two vertical axes with differing scales, two roughly parallel curves, and no legend to indicate which scale was for each curve. In the instrument, it was related to the proposition that "storks bring babies."

The second case made use of a graph plotting dietary fat intake versus deaths from breast cancer for a number of countries. The scatterplot, typical of a relatively strong positive correlation, is seen in Figure 11.2. This graph also was adapted from *Scientific American* (Cohen, 1987). It was related in the instrument to the proposition that "high intake of dietary fats help cause breast cancer." In the actual graph, there were over twice as many points, each labeled with a country name, as is Hong Kong in Figure 11.2. Breast cancer was mentioned specifically as the cause of death only in the original caption, which was part of the stimulus seen by subjects.

The third case used a graph presented in a monograph on the effects of being overweight (Mayer, 1968, p. 74), which has also been reprinted in a small monograph sometimes used in introductory sociology courses (Scitovsky, 1976, p. 157). It used a pair of graphs to relate occupations which demanding increasing degrees of physical activity to daily caloric intake and to body weight. The

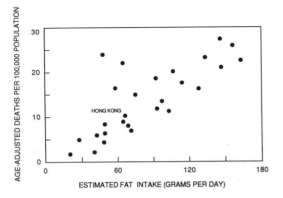

Figure 11.2. Age-adjusted deaths per 100,000 from breast cancer versus dietary fat intake. (Simplified graph adapted from Cohen, 1987.)

graph as simplified is presented in Figure 11.3. It was related to the proposition that "increasing levels of physical activity are associated with lowering body weight and lower food intake (in calories per day)." In the original graph, the occupations were specifically named.

The fourth and final case was included to explore, in part, the effects of graphical data based on a mathematical model used for extrapolation rather than based on empirical data from the past. The pair of graphs presenting the simple predictive model appeared in an advertisement placed by the Environmental Fund in the *Christian Science Monitor* of August 10, 1976 (as well as in other newspapers) and has been reprinted in at least one introductory text on sociology (Calhoun, 1978, p. 13). A simplified adaptation of these graphs presented in Figure 11.4 depict a projection of the growth of food production and population in developed and developing noncommunist countries. These graphs were related to the proposition that "food production will increase faster in developed countries than in developing countries." In the original, the percentages were imbedded in the body of the graphs, with small arrows pointing to the appropriate curve.

A rough taxonomy of information-processing tasks assumed *a priori* to be part of the critical evaluation of graphical arguments was used to structure a series of questions in the four cases—that is, related to the four graph(s)/proposition pairs. That taxonomy involved five levels of tasks as follows: (i) obser-

vation of facts in the graph(s), (ii) observation of relationships in the graph(s) as graphs, (iii) interpretation of relationships in the graph(s) in their "real-world" context, (iv) evaluation of the value of the graphical data as evidence for the truth of the related proposition, and (v) assessment of the basis on which each subject made his/her evaluation of the evidential value of the data (metacognitive awareness of critical evaluation).

One or more questions were presented in the printed instrument for each of the levels of the taxonomy and each of the cases. Some questions were multiple choice and other (especially interpretive) questions were open-ended.

The subjects

Seven adult subjects were used for this feasibility study (along with one high school student, the results for whom are not reported here). While this was largely a sample of convenience, some effort was made to introduce sources of variance in background, training, and approach into the pool of subjects. The differences of expert and novice performance have been widely used in studying problem

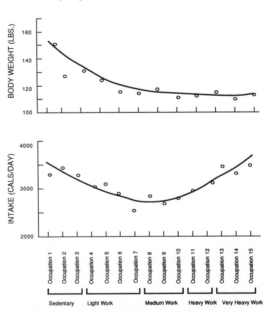

Figure 11.3. Caloric intake and body weight versus occupational activity level. (Simplified graph adapted from Mayer, 1968; also reprinted in Scitovsky, 1976.)

POPULATION AND FOOD PRODUCTION GROWTH

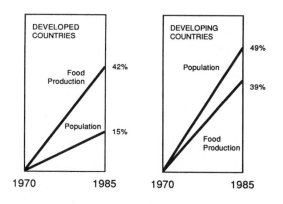

Figure 11.4. Food production and population growth in developed and developing countries. (Simplified graph adapted from advertisement by the Environmental Fund in *Christian Science Monitor,* August 10, 1976; reprinted in Calhoun, 1978.)

solving (perhaps beginning from the work reported in Larkin, McDermott, Simon, & Simon, 1980). A rough attempt was made to do this here by including among the subjects three university faculty and four graduate students just beginning their disciplinary training. If was further assumed that academic discipline would be a source of differences in approach, and an effort was made to obtain at least one subject in each of the two groups in mathematics (for the effect of specific training in at least certain categories of graphs), in a social science (for experience in critical evaluation of narrative reasoning as opposed to the more formal reasoning of mathematical proof), and in one of the humanities (to introduce still different traditions of reasoning and interpretation, presumably ones more naive about or at least less involved with quantitative elements). By deliberately using these two sources of variance, it was felt that even this small convenience sample could sketch some of the spectrum of response to the instrument that would be typical of much larger samples of adults (but probably not of younger, less developmentally advanced subjects).

The seven subjects will hereafter be known by a code as follows: (i) M1, a full professor of mathematics (in the area of analysis); (ii) M2, a graduate student in mathematics (topology) about to begin his doctoral

dissertation; (iii) S1, an associate professor of geography (whose particular research often used quantitative approaches; (iv) S2, a beginning graduate student in education; (v) S3, a more advanced graduate student in mathematics education; (vi) H1, an assistant professor of English (a specialist in Chaucer) whose work was totally in a nonquantitative tradition; and (vii) H2, a beginning graduate student in history whose training had been a more humanistic approach rather than a "scientific" approach to the study of history.

The procedure

Of particular interest in the present context are the methodologies of assessing individual competencies in the critical evaluation of graphical arguments. The survey procedure was designed with this in mind. Each subject was asked to complete the printed instrument, but in a "talk-aloud" format in which the subject was audiotaped. The instrument was completed in the presence of the researcher, but without interaction other than minimal prompts to remember to talk aloud. After the printed instrument had been completed in the talk-aloud format, the researcher conducted a brief follow-up interview with each subject to gather additional information based on his observation of the subject completing the printed instrument.

The responses to the printed instrument thus constituted one source of data. A second source of data was the transcript of the talk-aloud completion of the instrument. A third source of data, used more sparingly since it relied more on recollection and less on immediacy, was the follow-up interview transcript. Analyses of responses to the printed instrument could be contrasted to analyses based on the richer data sources from the interview. A rough assessment of the limitations and possibilities of getting at assessments of the targeted competencies through group paper-and-pencil testing could thus be made and compared to results from the more resource-intensive assessment methods of the interviews.

More complete cognitive task analyses of these data and more complete comparisons of the alternative assessment methodologies are planned for the future. The remainder of this chapter sketches in some of the more consistent findings for the tasks at each of the five levels of the taxonomy and points to some of the issues raised by these results.

Observation of Facts

Each of the four cases (sections related to one of the four graphs or sets of graphs) began with a simple question to explore the subject's ability to interpret graphical data sufficiently well to observe single facts from a study of the graphs. It was presumed, a priori, that such competencies in the observation and interpretation of simpler graphs were foundational for the more complex skills required for tasks that occurred later in the rough taxonomy of proposed tasks.

All subjects showed themselves to be reasonably strong at such observational tasks. The following were the lead questions for each of the four cases:

(i) Figure 11.1. "Approximately what were the number of newborn babies in West Germany in 1965?"

(ii) Figure 11.2. "Approximately how many age-adjusted deaths from breast cancer per 100,000 population did Hong Kong have according to the data presented in this graph?"

(iii) Figure 11.3. "Approximately what was the average food intake in calories per day for the average of the four occupations in the activity group, Light Work?"

(iv) Figure 11.4. "What is the projected food production in 1985 for the developed countries?"

Each of these questions was multiple choice with a "cannot tell from this graph" option (which was correct only in the case of Figure 11.4, since what was represented was only percentage change).

For Figure 11.3, all seven subjects answered correctly. For Figure 11.4, only subject H1 answered incorrectly. Her comment from the transcript at this point was "Is it a 42 percent increase?...I'm not very confident." For Figure 11.2, only subject H2 gave an incorrect answer. Her comment at this point was "How many age-adjusted deaths from breast cancer? OK. Per 100,000 of population. Well, without knowing the population of Hong Kong, you can't tell from the graph."

From this it was clear that this subject simply had not properly interpreted the units of the vertical axis of the graph in Figure 11.2. Failing to realize that it was already in deaths per 100,000, she assumed that she needed to know the overall population of Hong Kong to calculate the desired rate. Other questions show H2 to be a careful observer and this was a relatively rare slip which was not repeated in other, more complex situations.

The question for Figure 11.1 presented a more complex case. As presented in the instrument, the graph had two curves and two vertical scales with nothing to indicate which curve was for each scale. Further, the question asked subjects to read a point from the far left of each curve against the right-hand vertical scale. There was thus an extra complexity to the task, as well as an inherent ambiguity. Although among the choices given for answers were correct answers for each of the two curves, since the two curves were not labeled, the correct answer was a disjunction that combined the correct answer for each of the two curves and which occurred later in the list of choices.

In this context, four of the subjects (the three faculty members and H2) chose the first "correct" answer in the list, (b), which was for one of the curves and not the other. This was likely an artifact of the usual procedure of ceasing to pursue the choices in a multiple-choice item when a correct choice is encountered. These subjects thus likely never read the disjunction options (as was indicated by the transcript information). The other three subjects (all graduate students) went directly from the ambiguity of having two curves without a distinguishing legend to the final choice of not being able to tell from the graph. Essentially, none of the subjects considered the disjunction or engaged the possible answers fully. This appears, in context, to reflect an artifact of the questioning rather than a lack of interpretational skill. As an indicator of observational skill, however, it does suggest a tendency for these subjects towards superficiality in simple interpretive questions.

To summarize, all of the subjects at this level of development seemed quite skillful at interpreting even flawed and difficult graphs, making few errors other than not fully to engage the complexity of interpretation in Figure 11.1. This skill in observation of facts could easily be assessed through the use of the printed instrument only, although the supporting transcript data shed light on the nature of the few errors that occurred.

Observation of Relationships

The second kind of tasks postulated in the taxonomy were those of observing relationships within graphs, of interpreting the graphs only as a visual display without reference to the meanings of graphical elements in context. This sort of task was again explored, in part, by a series of multiple-choice questions, such as "Considering the two curves of the graph only as marks on a piece of paper, how do the changes in these two curves compare?" (for Figure 11.1) or "Considering this graph only as dots on a piece of paper, what pattern do these dots (points) on the graph reveal?" (for Figure 11.2). The choices in each of the four cases varied as was appropriate for the graph involved. For instance, the four choices for Figure 11.1 were: (a) For the most part, both curves tend to increase or decrease together, That is, as one shows a decrease from year to year, so does the other; (b) For the most part, one curve tends to increase as the other decreases, and vice versa; (c) For the most part, there is no relationship between the changes in the two curves; and (d) There is a relationship other than that in "a" or "b" above. Please specify.

In each case, the final option was that there was an "other" relationship than those listed in the choices, and the possibility of an open-ended response was provided.

The results again suggest that skills in the observation of relationships in graphical presentations were not particularly problematic for subjects at this developmental stage. For the questions related to Figures 11.2 and 11.4, all seven subjects chose correct answers. For the questions related to Figures 11.1 and 11.3, one subject in each case made an "error," at least in the sense of not selecting the most correct of the choices given in the list and offering instead a more problematic response.

For the question on Figure 11.1, subject M1 (the mathematics professor) chose "c,"

which rejected both of the stated relationships. The transcript reveals his rationale when he states, "There are small time intervals on which they do different things." It thus seems clear that M1 made no interpretive use of the phrase "for the most part" included in each of the listed responses. Although it is pure speculation, this may well have been because of the standards of precision of relationship typical of the mathematical study of the behavior of curves and function graphs. The background and training of M1 may well have made him less likely to appreciate approximate behavior in a way more typical for those trained in social science disciplines.

For the question on Figure 11.3, subject H2 again made an idiosyncratic response, although one that was correct but not typical of the responses of the other subjects. In the talk-aloud protocol, she says,

It's interesting. I'm trying to figure out the relationship because there is one, but it's not one that is suggested here.... There's not really a clear relationship between the two lines just as marks on the paper... which is what the question wants to know.

She then chose the option that allowed her to specify a relationship other than those given in the standard choices and wrote,

The relationship between the curves are different on the left half of the graph and the right half.

A look at Figure 11.3 reveals that this is a correct observation for one of the two graphs involved.

During the follow-up interview, clarification was sought for her response, and the subject stated that she chose to specify an "other" response rather than to choose the more common response ("There is a strong relationship like that suggested in 'a' [both curves tend to increase or decrease together] for the occupations in the 'Sedentary' and 'Light Work' activity groups.")

Because it didn't seem to say enough.... It's really similar but it seems to me to leave out the other half of the graph and I think there's a relationship there, too. It's just a different one.

I just didn't feel comfortable limiting myself to half of the graph.

Her response is equally correct and certainly more thoughtful than the more stereotypical responses of the other six subjects.

Observing relationships in graphical data thus seemed little more complex than observing facts in graphical presentations, at least for subjects at this developmental level. On the other hand, for the first time, at least a few differences began to emerge from the verbal report data that suggest processing differences due to differences in training and due to differences in personal cognitive style and background.

The paper-and-pencil assessment device using a multiple-choice format, even without the open-ended "other" possibility, would have captured virtually all of the responses that were largely stereotypical. However, as the excerpts from the transcripts reveal, there is a richness of evocation and insight provided by the open-ended response and the talk-aloud completions that is certainly valuable for research purposes. Until considerably more is known about the factors involved in observing relationships in graphical presentations, it seems inappropriate to go to more restrictive assessment procedures, even for this relatively less complex task. Certainly, this would be the case for investigation of such tasks with developmentally less mature subjects, for whom verbal report data seem essential.

Interpretation of Relationships

Assuming that the subjects were able to observe individual facts and relationships in graphically presented data, the next step in critical reasoning about this information would be the interpretation of relationships noted in terms of the substantive content of the situation depicted in the graph. Tasks of this sort thus formed the next level of the initial taxonomy.

Tasks in the survey instrument relative to this category took the form of open-ended questions that were appropriate variations of that for Figure 11.1, which was stated as follows:

Given your understanding of the relationship between changes in the two curves and given your understanding of what each curve represents, state in your own words in one or two sentences what fact is represented by the relationship between the two curves.

(In retrospect, it was realized that it would likely have been better to phrase the task instructions to include the verb "interpret" and to eliminate the noun "fact" from this context. However, data had already been collected using the task instructions as indicated.)

A survey of the resulting interpretations reveals statements of quite different types due to the nature of the interpretive task and, perhaps, in part, as an artifact of the already discussed flaws in the task instructions. Roughly, those interpretations were classified into one of four categories: (i) paraphrases, (ii) non-statements of relationship, (iii) statements of relationship, and (iv) inferences and technical statements.

By *paraphrases* is meant statements in which the task seems to have been taken as that of paraphrasing the graphically presented facts into English sentences, which resulted in descriptive statements of factual content with minimal interpretive elements. A clear example of this type of response is subject S3's response to the question for Figure 11.4:

In developed countries during the period from 1970–1985, food production increased 42% while population only increased by 15%. In developing countries during the same period, food production increased 38% while population increased by 49%.

This response constitutes nothing other than a catalog of facts drawn systematically from the graph and could serve as a written paragraph replacing the graph (hence, the phrase "paraphrase"). Such descriptive responses were rare but, when they did occur, suggested that minimal interpretive effort had been involved directed at *relationships*. This, however, may have been an artifact of the task instructions.

A second category has been labeled *non-statements of relationship*. Responses of this sort simply stated that there was a relationship without attempting to interpret or even describe the relationship. A typical example was the response of H2 to the task for Figure 11.4: "They indicate a relationship of speed of increase in food production versus population growth for these years in these kinds of countries." Such a statement asserts a relationship and categorizes it broadly, but does not attempt to state that relationship verbally. Another example is provided by M1's response to the same question: "There seems to be a relationship between increase in food production and increase in population." Responses of this sort, which occurred occasionally, should not be confused with the one response that might be called a "statement of nonrelationship" — "There is none" (M1's response to the question for Figure 11.3).

The third category identified was labeled *statements of relationship*. A typical example is the response of H1 to the task for Figure 11.4: "In developed countries food production outgrows population. In developing countries population grows faster." A second example is S2's response to the same question:

In developed countries food production is increasing faster than population. In developing countries population is increasing faster than food production.

Such statements capture the essence of the relationships exhibited in the graphical data and set them out in a relatively straightforward narrative manner. Clearly, interpretive processing has gone on, much more so than is evidenced by the example of paraphrase stated earlier. It should be noted that the differences between paraphrase and the present statements of relationship were not coupled with any noticeable differences in the quality of responses to tasks on observing facts and relationships in the graphs.

A fourth category involved interpretive statements that went beyond the mere state of relationships to draw *inferences* or to recast interpretations more in *technical statements* characteristic of a particular discipline. An example is provided by H1's response to Figure 11.3: "People who do heavier work can consume more calories and not weigh more." A second example is provided by S1's (the

geographer's) response to the task for Figure 11.4:

In developed countries, during the years represented, developed countries will have a greater food/population ratio in 1985 than they did in 1970. In developing countries, the will have a lower food/population ratio.

The categorization of responses to interpretation tasks is certainly only the beginning of attempts to characterize and understand the sorts of processing involved. Little further work has been done, other than to note two issues of some interest for further exploration.

First, although graphical data that do not involve time as a variable or represent the result of systematic variation are most appropriately described with the language of association, it was far more typical to find interpretations stated in the language of association. One example of what is meant by the language of variation is provided by H2's response to Figure 11.2: "For this data, larger fat intake and higher incidence of cancer occur at the same time in the same places." A second example is that of S2's response to the same graph, although this response focused on only part of the data: "In countries with high fat intake there is also a high rate of age-adjusted deaths from breast cancer."

In contrast, the language of variation is exemplified by S2's response to the task for Figure 11.4: "For sedentary and light work groups body weight decreases with decreased food intake. This relationship does not hold for groups after light work." This sort of variational language ("increasing," "decreasing") is by far the most typical in the responses gathered. It is worth further investigation to explore whether this is a simple verbal reflection of a mental movement through the data of the graph or graphs or whether it represents something more significant. Further, the two examples given for the language of association occurred for one of the more statistically naive subjects and one of the more statistically well-trained subjects. It seems worth further exploration to determine what leads to these language differences and whether there is one or more bases on which the choice for the language of association rather than for the language of variance is made.

A second fact that seems to emerge from these responses is that the level of detail and qualifications or limitations of interpretation seem to increase with doubt about the truth of the data or, at least, truth about the story that the data seem to be telling. Perhaps the best examples of this are seen in responses to the interpretive task for Figure 11.1, whose data on a relationship between storks and babies evoked early doubts in almost all subjects, even before the paired proposition was stated. Typical of the responses is that of M1 (italics introduced to indicate qualification and limitation of the statement of relationship): "The only relationship is that *over a 17-year period* both curves decrease. *However, over the next three years, this relationship does not exist.*" A second example is provided by the response of S1 to the same question: "*For the most part,* the two phenomena co-vary in time. However, *this does not indicate a clear causal relationship.*" It should be noted in the case of the last example that no statement related to cause and effect had been put forward in the instrument and, indeed, no statement of a related proposition had yet been offered. It is certainly worth investigating further the extent and nature of qualifying remarks in the presence of doubtful data. If consistent indicators could be found, it might be possible to further assess this aspect of thinking through open-ended printed instruments without relying on interview or talk-aloud data.

Evaluation of Evidential Value of Quantitative Information

Once relationships have been observed in graphical presentations of data and interpreted in terms of the substantive context of the content of the data, the next type of task in the taxonomy was to relate the interpreted data to the paired proposition, the truth of which is to be demonstrated, and to evaluate the value of the data as evidence for the truth of the proposition. Graphical data can be used merely as examples, as single pieces of evidence, as relatively conclusive pieces of evidence, as misleading evidence intended more to persuade than to legitimately prove, or even as cosmetic devices to make a discussion more interesting

with no evidential value intended. Certainly, individual evaluations of evidential value of graphical data should depend on the quality of the data and the relevance to the truth of the proposition. It may also be true that these evaluations depend, perhaps inappropriately, on belief in the truth of the proposition, since some may hold the implicit belief that even partial evidence cannot be offered to support the truth of a proposition believed to be false. The systematic variation in the cases for Figures 11.1, 11.2, and 11.3 allow for some exploration of these factors.

In each case, a task on the evaluation of the evidential value of the graphs involved was included by means of a multiple-choice question, the options for which varied as appropriate for the case. For instance, for Figure 11.1, the question and options were:

If this graph was offered as a piece of evidence to prove true the statement, 'Storks bring babies,' how would you describe the connection between the graph and the attempt to prove the statement true? (a) The graph is conclusive evidence for the truth of the statement; (b) There is no connection between the graph and proving the statement true; (c) The graph is consistent with the truth of the statement and thus might be considered one piece of evidence supporting the statement as true; (d) Connecting this graph to any argument for the truth of the statement would be an attempt deliberately to mislead those one is trying to convince.

For Figure 11.3, a fifth option was added, "A large part of the graph is consistent with the truth of the statement and thus might be considered one piece of evidence supporting a more restricted view of the statement as true." For Figure 11.4, which introduced graphical presentation of projections from a mathematical model, a different fifth option was added, "These graphs represent a model projecting into the future (at the time the data were gathered) but, to the extent that the graphs are a small simplification of actual data and represent only a slight variation from the facts, they suggest strongly that the statement is true."

Only for the question related to Figure 11.2 (in which a relatively clear graph supported a proposition widely believed to be true) was there any uniformity of response, with all seven subjects indicating that the graph was consistent with the proposition and might constitute one piece of evidence for its truth. In each of the other three cases, no more than three subjects chose the same option on each question. This certainly suggests the limited insight to be gained from paper-and-pencil assessment at this point, and the discussion that follows will proceed case by case, relying heavily on the verbal report data.

In the case of Figure 11.1, three subjects (M1, S3, H2) indicated that there was no connection between the graph and the truth of the proposition. M1 explicitly said, "As far as I can see there is no connection between the graph and proving the statement true." The verbal reports of the other two showed that they used a process of elimination to arrive at the same answer, but did not explicitly state that they saw no connection nor why they considered there to be no connection. Two subjects (H1, S2) indicated that the graph was deliberately misleading when offered as evidence for the truth of the proposition. On follow-up, S2 indicated that he made his choice "because of knowing that [the proposition] was not true." In the talk-aloud transcript, subject H1 reveals something of her thinking:

Well, I don't think they can use that graph for that [proving storks bring babies]. But is that just because I don't believe storks bring babies? If this were some other kind of question would I be convinced by that graph?... No, because I want the [right] end of [the graph] explained. OK. I'm going to circle "d" ["deliberately misleading"] because I don't think it can be that.

The transcript for H1 explicitly demonstrates greater metacognitive awareness than those of the other subjects (although such awareness may have been present in others, but not captured due to the implicit sampling of cognitive processes in talk-aloud data). Clearly, she has engaged the question of the extent to which she is swayed by the falsity of the proposition and arrived at an answer that is, to her satisfaction, independent of that factor.

The other two subjects (S1, M2) indicated that the graph was consistent with the truth of

the proposition and might be considered a piece of evidence in support of it. S1, the geography professor, indicated

I'm debating between "b" [no connection] and "c" [consistent, a piece of evidence]. My mind says there is no connection between the graph and the statement, but, on the other hand, the graph is not inconsistent with that effort [to prove the statement] in terms of its shapes and parallels....So what's hanging me up is the phrase 'thus might be considered one piece of evidence supporting the statement as true.' One could use,...if one really believed that and was trying to prove it, the graph would at least be consistent with that effort. So I would probably go with "c."

This was the choice indicated in writing on the printed instrument. The richness of metacognitive awareness certainly could be captured only through transcript data. Further, this transcript shows the deliberate effort to move from self-awareness ("What's hanging me up...") to generalization about evidence and proof apart from this specific proposition (notice the change of person and change of tense to the subjunctive to consider the hypothetical, "If one really believed...").

M2, the mathematics graduate student, largely used a process of elimination to remove three choices and then began to decide between "b" [no connection] and "c" [consistent, piece of evidence]. At this point, the transcript reveals him stating

I would answer [that] the graph is consistent with the statement because brooding storks, newborn babies, they're both declining generally [in the graph]. Whether the statement's true or false, at least that's consistent with the statement and it might be considered as a piece of evidence as long as it's not considered as a hard and fast proof.

This segment of M2's transcript is largely opaque relative to his self-awareness and attempts to generalize about evidence, but the final statement is revealing. His training as a mathematician has been in terms of "proof"; here, he moves to a different kind of argument that allows cumulative evidence rather than deductive ("hard-and-fast") proof. What seems to be involved here is personal style and background overcoming specific training to allow a versatility in what is considered an argument that would not be true for a mathematician who considered "proof" in a narrow, technical and formal sense.

Overall, then, in considering responses to Figure 11.1, three subjects moved quickly to the conclusion that there was no connection between graph and proposition—they seemed to not seriously engage the question, perhaps due to the obvious falsity of the proposition. One other subject seemed to hold the implicit belief that evidence presented in support of a false proposition must inherently be an attempt to mislead, while another felt the connection to be an effort to mislead because of her interpretation of the graph as not consistent with the proposition. Two subjects arrived at a seemingly more appropriate response that allowed evidence to be presented in support of a proposition they believed to be false. Two of the more expert subjects revealed explicit self-awareness of the difficulties involved, and one explicitly revealed the move to generalization to find criteria for decision not tied to the specific proposition. At least one revealed the effects of, perhaps even conflict of, formal training and personal approach.

In the case for Figure 11.3, the data were again revealing. Three subjects (M1, H2, S3, the same as for Figure 11.1) again found there to be no connection between the graphs and the truth of the proposition. Unlike Figure 11.1, in which the evidence presented was largely consistent with the truth of what was obviously a false proposition, in the case of Figure 11.3, the proposition was partly true and partly false and the evidence as a whole did not support the proposition, although a restricted part of the evidence would support part of the proposition (one graph supported one part of a disjunctive proposition). In this case, while S3 again revealed no explicit basis for his conclusion, M1 revealed that it was based on his interpretation of the data to include both parts of the proposition.

The response of H2 to Figure 11.3 was idiosyncratic, but interesting. The transcript reveals something of her thinking:

Increasing levels of physical activity is [sic]

associated with lowering body weight and lower food intake. I guess I would state that [the graph] is consistent with the statement. Ah ha. Wait a second. Increasing levels of physical activities... hmm ... No, I'm not going to say that. It's not associated with lowering body weight. It is associated with a lower body weight. It doesn't indicate how the body weight changes. That's probably quibbling but I vote for "b" [no connection].

In fact, of course, H2 is quite correct in her quibble based on an unintended miswording in the instrument, but one which caused this subject, perhaps because of her disciplinary training that included sensitivity to nuance in language, to engage a different proposition than did the others and arrive at a conclusion appropriate to the proposition engaged. In sum, two of these three subjects moved directly from the falsity of the evidence to a conclusion of "no connection" as if implicitly holding the belief that "bad evidence cannot support even a true statement," while the third engaged effectively a different proposition due to interpretational differences.

Three subjects (S1, S2, H1) arrived at the conclusion that this graph was deliberately misleading as evidence for the proposition. The transcripts for all three subjects revealed that they explicitly went from deciding the data contradicted the truth of the proposition to concluding that the connection of this contradictory data to an attempt to prove the proposition true was misleading. They seemed clearly to believe, perhaps correctly, that presenting evidence that proves a proposition false as part of the case for its truth is a deliberate attempt to mislead. The details of the transcripts are not presented here for the sake of space, but all explicitly engaged the entirety of both graphs and both parts of the proposition, and thus correctly arrived at the contradiction between data and statement.

Only one subject (M2, the mathematics graduate student) arrived at the conclusion that a "large part" of the data were consistent with and evidential for a "more restricted" version of the proposition. The versatility and willingness to move from whole to part is consistent with his willingness to allow different canons of "proof" as discussed earlier, and his reasoning in doing so is revealing:

A large part is consistent thus might be considered one piece of evidence supporting the statement. Ah, "e" ["deliberately misleading"] again,... I think it would be foolish just to throw out the data... With more information, with better information, as I said, like having these heavy workers and very heavy workers like double their intake up to seven and eight thousand and seeing what happens then to body weight, that might come up with a more conclusive statement.

Certainly, the data of the graph seem to constitute a much more concrete picture for this subject in which he can imagine experimentally varying the factors and in which he is willing to consider the consistent parts of evidence and proposition and suggest further exploration of the more contradictory parts.

Overall, in relation to Figure 11.3, two subjects went from perceiving the evidence as not supporting the claim to a conclusion of "no connections" which suggests again a failure to actively engage questions of evidential value. Three others explicitly went from determining the full data to contradict the full proposition and thus to a conclusion of an attempt to deliberately mislead. The last subject discussed made a move that the others did not consider or consider legitimate, to consider part, rather than all, of both data and evidence. This led him to a different conclusion consistent with his approach to the task.

Evaluating evidential value of graphical data, it would seem, depends on both the perception and interpretation of the data to serve as evidence, and on the perceived truth of the proposition to which the data are to be linked as evidence. When the proposition is perceived as likely true and the evidence is readily interpretable as consistent with this (as in Figure 11.2), there was a uniform assignment of positive evidential value to the data. When the data were perceived as disproving the proposition (Figure 11.3, for all but M2 who moved from the whole to the part) — as being reasonably definitive evidence of its falsity — the presentation of such data in support of the truth of the proposition was relatively often seen, appropriately perhaps, as deliberately misleading.

When the proposition appeared false (the case of Figure 11.1), but the data would, in general terms, possess some positive evidential value, the responses were more complex. In the case of both Figures 11.1 and 11.3, some subjects failed to engage in any serious way the questions of evidential value going from any element of falsity in either data or proposition to the conclusion that there was no connection. However, among those who seriously engaged the question of evidential value, some could not escape the specifics of the false proposition and seemed to respond based on an implicit belief that presenting evidence in support of a false proposition was deliberately (inherently?) misleading. Only a few subjects were able to generalize the question of evidential value beyond the specifics of the particular case and conclude that, although they disbelieved the proposition, the data still have (albeit limited) positive evidential value. The case for Figure 11.4, not discussed here, further supports the sort of result seen for Figure 11.1.

Certainly, tasks of evaluating evidential value of graphical data have been seen here to be more complex than the earlier observational tasks. Clearly, a variety of factors make an impact on the performance of such tasks: disciplinary training, personal cognitive style, metacognitive awareness, ability to generalize, and so on. Such tasks are worthy of further serious research. It also seems clear from the preceding discussion that the printed instrument does not capture data rich enough to allow productive research in this area, and that some form of verbal report data is demanded. It is, of course, possible that as further research develops more complete cognitive task analyses of these sorts of tasks, paper-and-pencil assessment can be made more revealing.

Assessment of the Evaluation of Evidential Value

The final level of tasks postulated on the taxonomy for critical evaluation of quantitative arguments is that of assessing one's own evaluation of the evidential value of quantitative data used in support of a stated proposition. It is assumed that such metacognitive self-awareness and self-criticism is an expected part of the exercise, growth, and development of critical thinking of any sort and that, formally or informally, such activity accompanies efforts to critically evaluate any quantitative or graphical aspects of an argument. Because of space considerations, only a brief sketch will be offered of findings from the survey in this area.

This kind of task was explored initially through multiple-choice questions for each case which asked subjects, based on whether they found the graphical evidence convincing or unconvincing, to indicate a reason why the graph was unconvincing or convincing "as proof that the statement was true." Typical choices for being unconvinced included that "It provided no information relevant to proving the statement true"; "The data were wrong"; "I already knew that the statement was false"; "Some apparent relationships happen just by chance"; and "The data presented in the graph were too limited to conclusively prove the statement true." Reasons given for being convinced were positive counterparts of these statements, with slight variations in wording from case to case.

When the graphical data were clear and the proposition relatively well accepted as true, there was considerable uniformity of response. For example, in the case of Figure 11.2, the data on diet and cancer, all subjects indicated that they were convinced rather than unconvinced by the data. Further, all chose the same primary reason (subjects were allowed to designate a primary and a secondary reason from among the choices), that the information provided was relevant to proving the statement true. The transcript data revealed some diversity in how individual subjects assessed relevance, and further work in this area certainly requires analysis of such criteria. Until such work is done, "relevant" must remain something of an ambiguous or "weasel" word in this context. It expresses a feeling, a reaction, but the basis for that reaction needs further elucidation through interview studies.

When the graphical data clearly contradicted the truth of the proposition, virtually all subjects were unconvinced by the data and indicated that this was because "the graphs demonstrate that the statement is not true." This was the case for six of the seven subjects

for the task for Figure 11.3. The seventh subject was M2, who moved from attending to the whole of the graph and proposition to whether part of the graph supported one part of the disjunctive proposition. He, too, indicated being unconvinced, but indicated as a reason that "the data presented in the graph were too limited to conclusively prove the statement true." This response is somewhat unsurprising in light of earlier transcript segments quoted for M2, in which he described the further data he felt necessary to collect to clarify the part of the graph and proposition that he felt were inconclusive.

When the falsity of the proposition seemed certain, even though the graphical evidence was consistent with its truth, more diversity of response was obtained. This happened in the case of Figure 11.1. In this case, all were unconvinced, but the reasons indicated varied. Three subjects indicated that the graph "provided no information relevant to proving the statement true," three indicated that they were unconvinced because they "already knew the statement was false." Only one subject, M2 again, indicated that he was unconvinced because "the data presented in the graph were too limited to conclusively prove the statement true." This suggests that, even with adult subjects, there were varying degrees of self-awareness and ability to distance one's self from the particulars to consider evidential value in general. Even though two of the subjects' transcripts showed some metacognitive awareness of this need for distance (S1 and H1), their final assessment of their reactions were that they were influenced by their prior beliefs to the point that they were unconvinced. Clearly, there are strong interactions among factors such as prior beliefs, self-awareness, and knowledge of and training in the canons of evidence. It should prove interesting to explore these reactions for less-developed and less academically trained subjects, to see if there is greater uniformity of response, as well as lower self-awareness.

The case for Figure 11.4 elicited the most mixed responses, with two subjects (S1 and H2) convinced by the data, while the other five found the data unconvincing. Five of the subjects (the two who were convinced and three of the others) had as either their primary or secondary criterion that of relevance. Some found the information relevant, others not so. This points out even more markedly the ambiguity of criteria for relevance and the need to explore this area further. Three of the subjects gave as their primary reason a choice particular to the task for Figure 11.4, "Models and projections are not real data and cannot help to prove anything." The inclusion of this frankly exploratory choice in the instrument has led to results that suggest that the status of models and of predictive methods are a special category of evidence in the minds of many people, even those who are cognitively developed and academically trained, and are a subject that needs considerable further direct investigation. The data and graphs in Figure 11.4 appeared as advertisements intended to persuade readers about social issues in several national publications and are not atypical of other such efforts. If, indeed, many persons are inherently unconvinced by projections and by results drawn from mathematical models, this suggests a variety of instructional tasks that might confront mathematics educators, as well as some changes of policy on the part of those who seek to persuade the public on policy matters.

Concluding Remarks

As has been stressed from the beginning of this chapter, the data that served as the basis for much of this discussion served, at best, as a feasibility study to assess the value and possibilities of further research in the area of critical evaluation of quantitative arguments. The use of such case study methodology and transcript data may have caused the reader to be lost in a sea of details. To help end this discussion, a number of summarizing conclusions, as well as some suggestions for what implications this study might have for potentially useful future research, follow.

First, it should be noted that while the difficulty of observing facts and relationships in graphical data seems likely to vary with cognitive developmental level, for these adult subjects, these tasks were largely not problematic. They represented cognitive tasks well within the competence of these individuals, even when the approach (and, at times, the results)

varied from subject to subject.

Second, the interpretation of relationships seen in graphically presented data seemed a more complex and problematic matter, even for these subjects. A variety of levels of interpretation seemed to emerge (e.g., paraphrase, statements, inferences). The level at which interpretive statements had qualifying and limiting language seemed to increase for data that were more controversial or seemed to make a case for a proposition believed not to be true. The use of the rhetoric of association versus the rhetoric of variance emerged as a topic in need of more detailed study in future research.

Third, evaluation of the evidential value of graphical data also seemed a complex and problematic task. Some subjects seemed incapable of engaging the questions seriously, perhaps because they could not disregard prior beliefs in relation to the proposition under discussion. Different levels of metacognitive self-awareness were explicitly demonstrated and were seen to have implications for the accuracy of judgments about evidential value. Factors of training, background, cognitive style, and versatility (e.g., part/whole moves, alternate canons of proof) were seen to be involved. It certainly seems worthwhile to study further subjects' knowledge of one or more sets of canons of proof and evidence, as well as to probe through interviews to elucidate beliefs regarding the naive canons of proof held by individuals at different developmental levels and differing levels and types of academic training.

Fourth, while the look at how subjects assessed their own evaluations of evidential value of graphical data was more limited here, it certainly seemed sufficient to suggest the importance of the role of prior beliefs, both about what constitutes evidence and about the substance of propositions. Further, clear differences in self-awareness and self-criticism were noted, and it seems worthwhile to try to relate these to other metacognitive issues in further research. Finally, the special case of models and predictions emerged as an area in need of direct, special investigation.

In addition to this summary of what emerged from consideration of tasks from the five levels of the taxonomy of tasks in critical evaluation of quantitative arguments, a number of more general conclusions about these topics as an area of potential research can be stated. First, it seems that these questions are relevant to societal and educational concerns, that they are interesting and pressing, and that there are appropriate research methodologies by which they can be investigated.

Second, it seems clear that before paper-and-pencil assessment instruments become usable for research and evaluation in this area, that the more complex tasks of research with verbal report and interview data first must be undertaken. Rich sources of data must be used to provide the characterizations of cognitive processing that may allow eventually for much simpler assessment techniques.

Third, it seems that the kinds of tasks must be identified more clearly, and the initial rough taxonomy of task types greatly increased. Cognitive task analyses are needed of relevant tasks, as is more systematic exploitation of relevant research literatures (e.g., that on metacognition).

Fourth and finally, the tasks for the research agenda of mathematics education in this area have barely begun to be identified. The survey on which the present discussion was based was limited both in terms of the developmental levels of the subjects used and in terms of the quantitative aspects of arguments investigated (i.e., graphically presented data related to the truth value of single propositions). Certainly, it remains to extend these investigations over a larger range of the developmental spectrum and to relate findings to the literature on the relationship of mathematical problem-solving abilities to levels of cognitive development. Further, it remains also to extend these investigations to a larger range of quantitative argument types (e.g., statistical data, patterns in tables, critiques of entire arguments). Until these extensions are begun, the surface of a promising and needed area of research in higher order thinking with an inherently mathematical aspect will have been barely scratched.

References

Applebee, A. N. (1984). Writing and reasoning. *Review of Educational Research, 54*(4), 577-596.

Brown, A. L. (1975). The development of memory: Knowing, knowing about knowing, and knowing how to know. In H. W. Reese (Ed.), *Advances in child development and behavior*, Vol. 10, New York: Academic Press.

Calhoun, D. W. (1978). *Social science in an age of change* (2nd edition). New York: Harper and Row.

Cohen, L. A. (1987). Diet and cancer. *Scientific American, 257*(5), 42-48.

Ericsson, K. A., & Simon, H. A. (1984). *Protocol analysis: Verbal reports as data*. Cambridge, MA: The MIT Press.

Flavell, J. (1976). Metacognitive aspects of problem solving. In *The nature of intelligence*. Hillsdale, NJ: Lawrence Erlbaum Associates.

Franks, J. J., Bransford, J. D., & Auble, P. M. (1982). The activation and utilization of knowledge. In C. R. Puff (Ed.), *Handbook of research methods in human memory and cognition*, New York: Academic Press.

Gardner, M. K. (1976). Cognitive psychological approaches to instructional task analysis. In D. Klahr (Ed.), *Cognition and instruction*, Hillsdale, NJ: Lawrence Erlbaum Associates.

Glasman, N. S., Koff, R. H., & Spiers, H. (1984). Preface. *Review of Educational Research, 54*(4), 461-471.

Greeno, J. G. (1978). Nature of problem-solving activities. In W.K. Estes (Ed.), *Handbook of learning and cognitive processes, 5: Human information processing*, Hillsdale, NJ: Lawrence Erlbaum Associates.

Gregg, L. W. (1976). Methods and models for task analysis in instructional design. In D. Klahr (Ed.), *Cognition and instruction*, Hillsdale, NJ: Lawrence Erlbaum Associates.

Haney, W. (1984). Testing reasoning and reasoning about testing. *Review of Educational Research, 54*(4), 597-654.

Kail, R. V., & Bisanz, J. (1982). Cognitive strategies. In C.R. Puff (Ed.), *Handbook of research methods in human memory and cognition*, New York: Academic Press.

Larkin, J., McDermott, J., Simon, D., & Simon, H. A. (1980). Expert and novice performance in solving physics problems. *Science, 208*, 1335-1342.

Lefferts, R. (1981). *Elements of graphics: How to prepare charts and graphs for effective reports*. New York: Harper and Row.

Lester, F. (1982). *Mathematical problem solving: Issues in research*. Philadelphia: Franklin Institute Press.

Mayer, J. (1968). *Overweight: Causes, costs, and control*. Englewood Cliffs, NJ: Prentice-Hall.

Mayer, R. E. (1983). *Thinking, problem solving, and cognition*. New York: W. H. Freeman and Co.

Mayer, R. E., Larkin, J. H., & Kadane, J. B. (1976). A cognitive analysis of mathematical problem-solving ability. In D. Klahr (Ed.), *Cognition and instruction*, Hillsdale, NJ: Lawrence Erlbaum Associates.

National Council of Teachers of Mathematics. (1989). *Curriculum and evaluation standards for school mathematics*. Reston, VA: National Council of Teachers of Mathematics.

Polson, P., & Jeffries, R. (1982). Problem solving as search and understanding. In R.J. Sternberg (Ed.), *Advances in the psychology of human intelligence*, Vol. 1, Hillsdale, NJ: Lawrence Erlbaum Associates.

Resnick, L. B. (1976). Task analysis in instructional design: Some cases from mathematics. In D. Klahr (Ed.), *Cognition and instruction*, Hillsdale, NJ: Lawrence Erlbaum Associates.

Rowe, H. A. H. (1985). *Problem solving and intelligence*. Hillsdale, NJ: Lawrence Erlbaum Associates.

Schoenfeld, A. H. (1985). *Mathematical problem solving*. San Diego: Academic Press, Inc.

Scitovsky, T. (1976). *The joyless economy: An inquiry into human satisfaction and consumer dissatisfaction*. Oxford, UK: Oxford University Press.

Siegler, R. S. (1978). The origins of scientific reasoning. In R. S. Siegler (Ed.), *Children's thinking: What develops?*, Hillsdale, NJ: Lawrence Erlbaum Associates.

Simon, H. A. (1978). Information-processing theory of human problem solving. In W.K. Estes (Ed.) *Handbook of learning and cognitive processes, 5: Human information processing*, Hillsdale, NJ: Lawrence Erlbaum Associates.

12

Investigation of Structured Problem-Solving Items

MARK WILSON

The ultimate aim of measurement is to gain "objectivity and simplicity" (Glaser et al., 1987, p. 64) in relating real-world observations to theory. The concept of levels of understanding may be a useful intellectual tool in this process in a range of different situations. When a developmental continuum is postulated and found, levels may provide a convenient way to label portions of the continuum for practical purposes, or may be used to give practical expression to a researcher's doubts about the precision of measurement (i.e., as an alternative to simply reporting standard error of measurement). At the other extreme, as in much qualitative research, researchers may be striving to preserve the full impact of individual differences in development by avoiding standardization of the measurement process.

Nevertheless, the description of some sort of conceptual structure becomes all but inevitable as the researcher proceeds to interpretation. This structure may be more complex than a sequence of developmental levels; often a structure like a (mathematical) lattice will seem more appropriate. Here a sequence of developmental levels may prove useful if considered as a projection of this more complicated structure onto a single dimension. In between these two extremes lie the approaches that are based explicitly on developmental levels, such as the SOLO Taxonomy described in the following section.

The SOLO Taxonomy

The Structure of the Learning Outcome Taxonomy (SOLO; Biggs & Collis, 1982) is a method of classifying learner responses according to the structure of the response elements. The taxonomy consists of five levels of response structure: (i) a *prestructural* response is one that consists only of irrelevant information; (ii) a *unistructural* response is one that includes only one relevant piece of information from the stimulus; (iii) a *multistructural* response is one that includes several relevant pieces of information from the stimulus; (iv) a *relational* response is one that integrates all relevant pieces of information from the stimulus; and (v) an *extended abstract* response is one that not only includes all relevant pieces of information, but also extends the response to integrate relevant pieces of information not in the stimulus. This taxonomy is described in more detail by Romberg, Zarinnia, and Collis (see chapter 2, this volume).

In the particular type of SOLO item under study (Collis & Davey, 1986), a short piece of stimulus material which might consist of text, tables, or figures is supplied, then students are asked to answer open-ended questions concerning the material. Together, the stimulus material and the questions are referred to as a "superitem" (Cureton, 1965). An example is given in Figure 12.1 (see page

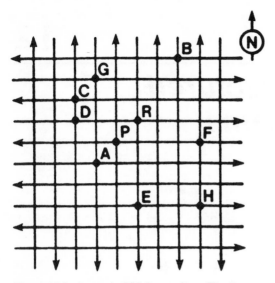

Figure 12.1. A sample SOLO superitem. The lines on the graph are city streets. One-way streets for vehicles are indicated by arrows.

failure on the lower level becomes not so exceptionable. To make this interpretation, however, one needs to go beyond the classification of responses into discrete ordinal classes and allow a probabilistic interpretation mapping response patterns into levels of the SOLO taxonomy, allowing for some range in the difficulties of questions that represent particular levels and allowing for the possibility of transitional states when classification may be difficult (Biggs & Collis, 1982, p. 174). This amounts to an Item Response Theory (IRT) interpretation, which is in line with the "hierarchical and cumulative" (Collis, 1983, p. 7) nature of the SOLO taxonomy. Its advantages are (i) the difficulty metric provides a means of directly comparing the ability of the learner with the items representing the SOLO taxonomy; (ii) the probabilistic formulation constitutes a framework within which response patterns other than Guttman true-types can be interpreted; and (iii) the probabilistic interpretation provides a scale of acceptability for response patterns, from those which are very consistent with the SOLO taxonomy to those which are very inconsistent with the taxonomy.

In the search for simplicity, a sequence of developmental levels can assist by allowing the classification of individual differences into two importantly different types. First, there are those individuals whose behavior is consistent with the sequence of levels, and who can therefore be assigned to an estimated position on the sequence. Second, there are those whose behavior is inconsistent with the sequence, the interpretation of which requires theory beyond that establishing the sequence. Clearly, the results of such a classification have important consequences for both theoretical research and individual measurement.

This chapter takes a set of mathematics problem-solving items that have been developed using the SOLO structure and subjects them to a detailed empirical scrutiny to check for empirical consistency with the levels construct. The fit of the empirical data to a formal measurement model is interpreted, and the relationship between a mathematics attitude variable and the levels construct is investigated. These are only a few of the empirical checks that could be made with these sorts of data. A paper by Wilson and Iventosh

199 for the complete text of questions). Each question is linked to a successive level of the taxonomy, starting with the unistructural level. Responses to each question are judged as acceptable or unacceptable according to an agreed set of criteria.

Because of the hierarchical nature of the taxonomy, it is expected that the majority of responses will be in the form of "Guttman true-types" (Guttman, 1941), where success on a higher level item can only be observed following success on the item at the level immediately below. That is, for most learners, success on one level of the taxonomy will be preceded by success on all the lower levels. Clearly, when the response to a superitem conforms to one of these patterns, it can easily be assigned to one of the SOLO levels. However, patterns other than these Guttman true-types can also arise, as has been discussed by Andrich (1985). For example, a learner might succeed on the unistructural and relational questions, but fail on the multistructural. One potential interpretation for a non-Guttman true-type response pattern such as this is to surmise that the relational question is relatively easy within the set of possible relational questions and that the multistructural one is relatively hard within the set of possible multistructural questions. Then if the two sets overlapped somewhat in difficulty, a response pattern of success on the higher level and

(1988) examined the consistency between the questions used to measure individual levels and the overall score across levels, and also explored the usefulness of a formal model to screen for students whose performance is not consistent with the levels construct.

The Partial Credit Model[1]

Although standardized achievement tests usually are composed of items scored right or wrong, there are many areas of educational achievement in which right/wrong scoring is considered inappropriate. It would not usually be considered appropriate to describe a performance in music, dance, or speaking as either "right" or "wrong," for example. Nor would it be considered appropriate to describe a piece of art, an essay, a technical drawing, or an industrial arts model constructed by a student as "right" or "wrong." When evaluating achievement in these areas, it is conventional to assign varying degrees of credit to students' attempts (e.g., by grading them on a scale of 1 to 5).

Even test items usually scored right/wrong can sometimes be scored in more than two outcome categories. Among the "wrong" (and sometimes the "right") answers that students give to an item it is sometimes possible to identify different kinds and levels of understanding. If students' answers to an item can be grouped according to the level of understanding that they reflect, then more than two ordered levels of outcome on that item might be defined.

The Partial Credit Model (PCM) (Masters, 1982) is a statistical model developed to supervise the construction and *measurement* of variables in behavioral and social science research. It belongs to the general class of models known as item response or latent trait models and, more specifically, to the family of Rasch (1960) measurement models (Wright & Stone, 1979).

The starting point in the use of the PCM is the intention to construct a variable along which individuals can be positioned. The next task is to develop questions, items, or opportunities that will provide relevant observations concerning this variable. Frequently, these observations will take the form of records of right and wrong answers to test questions. But the kinds of observations for which the PCM has been developed are based on more than two ordered response alternatives, such as SOLO superitems or the ordered rating systems exemplified in the work of Carpenter and Moser (1984).

The PCM can be thought of as an extension of or alternative to statistical models developed to supervise the measurement of achievement or competence from test items scored right or wrong. The observations used to construct measures of achievement with the PCM are, in general, very different from records of right and wrong answers; they are based on inferences about students' *levels* of understanding or competence.

The PCM provides a framework for assessing the validity of attempting to summarize performances on different aspects of achievement in a single global measure. Tests of item fit identify individual items which function differently from other items and may lead to the conclusion that it is inappropriate to attempt to summarize all aspects of competence in a single measure. The PCM can also be used to construct a "map" to show how students' understandings of a phenomenon change as they develop competence. In addition, the PCM provides a framework for identifying aspects of achievement in which a student is experiencing difficulty or making unexpectedly slow progress.

The PCM takes as its basic observation the number of steps that a person has made beyond the lowest performance level. Consequently, the parameter to be estimated is the "step difficulty" within each item. These step difficulties are substituted into the model equation for the PCM (Masters, 1982) to give a set of model probabilities for any given value of person ability. Figure 12.2 shows a plot of these model probabilities in a figure called an "item response map." The example shown is actually the item response map for the first item to be examined. Responses to this item have been scored in four ordered categories labelled 0 to 3. In this picture, ability increases to the right on the page from -4.0 to $+4.0$

---

1 Portions of this section and later sections of the introduction are adapted from Masters and Wilson (1988a).

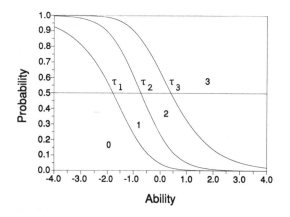

Figure 12.2. Item response map showing Thurstone thresholds.

logits. The logit scale is a log odds scale. Thus, for a dichotomous item, the odds of success are calculated by taking the antilog (to base e) of the logit difference, and the probability of success is found by solving the equation

$$L = \ln(P/(1-P)),$$

where P is probability, L is the logit, and ln is the natural logarithm. For a person who is 1.0 logits above an item, the odds of that person succeeding on the item is $\exp(1) = 2.72$, and the probability of success is $\exp(1)/(1+\exp(1)) = 0.73$. This calculation is useful for gaining a "feel" for the interpretation of distance in the logit metric, but it must be emphasized that the interpretation for polytomous items is somewhat more complex. The best strategy is to read the probability directly from the figure, as is done in the next paragraph.

From Figure 12.2, it can be seen that a person with an estimated ability of 0.0 logits (middle of the picture) has estimated model probability of about .05 of scoring 0 on this item; .18 of scoring 1; .42 of scoring 2; and .35 of scoring 3. The relative values of these model probabilities change with increasing ability so that, over the portion of the ability variable shown here, low scores of 0 and 1 become decreasingly likely, and a score of 2 on this item becomes increasingly likely up to an ability level of about 0 logits. As ability increases above this level, a score of 2 becomes less likely as the highest possible score of 3 on this item becomes an increasingly probable result.

The item response map in Figure 12.2 can be used to illustrate several important features of the PCM. Consider the horizontal line through the middle of the picture at probability P = .5.

The intersection points of this straight line, labelled here τ_1, τ_2, and τ_3, are known in the psychometric literature as "thresholds." In dichotomously scored items, there is only one threshold (or difficulty) for each item, defined as the position on the continuum at which the single ogive for that item intersects P = .5. Although thresholds defined in this way had been widely used in psychophysics (e.g., Urban, 1908) and biometrics (e.g., Aitchison & Silvey, 1957), it appears to have been Thurstone who first used this approach to associate regions of attitude and ability variables with ordered response categories. Thurstone referred to τ_2, and τ_3, as the "upper and lower boundaries" of category 2, and the difference $(\tau_3 - \tau_2)$ as the "estimated width of category 2 on the psychological continuum" (Edwards & Thurstone, 1952, pp. 173–174). For this reason, we refer to the intersection points shown in Figure 12.2 as "Thurstone thresholds." More recently, Thurstone thresholds have been incorporated into some item response theory models for ordered categories (e.g., Samejima, 1969).

One practical difficulty that arises in examining item response maps is that it is difficult to arrange more than two of them side-by-side in a reasonably sized figure. This is often required, as the items are most often interpreted in relation to one another. Thurstone thresholds provide a way to summarize information about several partial-credit items; simply place the Thurstone thresholds next to one another on a graph. A certain amount of detail is lost (in fact, information is provided only about the points at which successive cumulative probabilities reach .5), but this is always the case with a summary, and should not be a problem if the item response maps are provided as well. The Thurstone thresholds can be interpreted as the crest of a wave of predominance of successive dichotomous segments of the set of levels. For example, τ_1 is the estimated point at which levels 1, 2, 3, and 4 become more likely than level 0, τ_2 is the estimated point at which levels 0 and 1 become more likely than levels

2, 3, and 4, and so on. Thus, in the case of SOLO superitems, τ_1 is where it becomes more probable that a response will be above prestructural rather than prestructural, τ_2 is where it becomes more probable that a response will be above unistructural rather than pre- or unistructural, τ_3 is where it becomes more probable that a response will be multistructural or above rather than unistructural or below, and τ_4 is where it becomes more probable that a response will be extended abstract rather than multistructural or below.

The PCM makes no assumptions about the unconditional distributions of the persons along the latent trait, but does assume that the model adequately fits the data. Model fit can be assessed using a measure of fit called the "Item Fit t" for items and the "Person fit t" for persons (Wright & Masters, 1982), which is a transformed mean square statistic. The distribution of this statistic is not precisely standard normal, so it will be used to focus attention on the more serious problems rather than to make a strict decision about whether persons or items fit or not. For items, it is possible to find the empirical item response map, which allows visual inspection of items that have been selected on the basis of the Item Fit t as questionable. Another way to assess fit is to divide the sample of persons into groups with interesting and interpretable differences, re-estimate the parameters in each case, and examine the differences. Only if the model fits in the different groups can meaningful comparisons be made. These comparisons can be organized by using a statistic called the "standardized difference" between the estimates (Wright & Masters, 1982, p. 115).

Method

The items

The items used here were designed according to the SOLO superitem format. The seven items examined were originally part of a much larger study of the usefulness of the SOLO superitem format for assessment of mathematics ability (Romberg, Collis, et al., 1982; Romberg, Jurdack, et al., 1982). The superitems themselves are presented on pages 199–

203. Due to the age of some of the students, only the first four levels (i.e., excluding extended abstract) were assessed. In discussing the results, individual items within a superitem will be referred to as "questions" to help clarify the distinction between levels. The students were also administered a series of questions concerning their attitudes towards mathematics problem solving and mathematics in general.

The sample

The students were in classes ranging from grade 7 to grade 12, primarily grades 9 and 10. The classes were those participating in an evaluation study of a statistics curriculum, and consisted of two experimental groups (25 classes) and a control group recruited from among co-workers of teachers in the first two groups (10 classes) in the states of Wisconsin and Connecticut (Webb, Day, et al., 1988). This sample design is entirely idiosyncratic and should be seen, for the purposes of this discussion, merely as a way of collecting a fairly large group of students whose educational experience is relevant and varied with respect to the problem-solving items. The problem-solving items were used as part of a pre-test of the students; therefore, no details of the experimental design are relevant. In all, 1238 students with complete data on the seven problem-solving items were available for the analysis.

Results

The problem-solving ability variable

The distribution of the 1238 students along the latent variable defined by the seven problem-solving items is shown in Figure 12.3, where ability has been estimated using the PCM (and is expressed in logit units). Most of the students (90 percent) were estimated to be between $-.63$ logits and 2.38 logits (scores of 9 to 18). Thus, in interpreting the item response maps, attention will be focused on this portion of the ability scale. Note also the nonlinear relationship between the logit scale and the scores, which is indicated by the selection of score locations given below the figure. This will also need to be kept in mind when interpreting the item response maps. The analysis

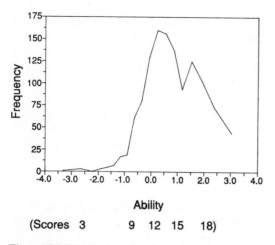

(Scores 3   9 12 15   18)

Figure 12.3. Distribution of students along the problem-solving items.

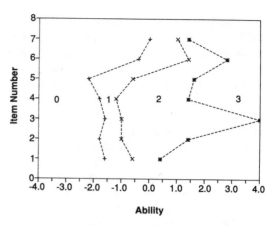

Figure 12.4. Thurstone thresholds for the problem-solving items.

was performed using the PC-CREDIT program (Masters & Wilson, 1988b).

Item 1 was shown in Figure 12.1, and the estimated item response map for this item was displayed in Figure 12.2. The conditional probabilities of response indicate that most students are performing above the prestructural level, ranging from approximately 15% of students with an item score of 0 (prestructural) at 9 points total (−.63 logits), to approximately 85% of students with an item score of 3 (relational) at 18 points total (2.38 logits). Thus, the predicted responses to this item range over the full SOLO spectrum within the range of ability of the majority of students in the sample. Moreover, the progress within the SOLO levels is quite regular from prestructural to relational for this item.

The relationship of the steps of item 1 to the steps of the other items is displayed in Figure 12.4, which records the Thurstone thresholds for all seven of the problem-solving items. In this figure, the unistructural threshold is marked by a " +," the multistructural threshold is marked by an "X," and the relational threshold is marked by an "*." Item 2, which concerns train timetable reading (see pages 199–200), has a similar pattern of thresholds to item 1 for the first two steps, but clearly has a more difficult transition to the relational level. The effect of this on the item response map for item 2 can be seen in the upper panel of Figure 12.5, where the band for score 2 (multistructural) is about twice as wide

as that for item 1. Moreover, the relational threshold for item 1 is the easiest to explain of all the items. It is interesting to consider the differences between the two "relational" tasks in attempting to understand this discrepancy. For item 1, the multistructural question requires the student to compare the distances from three separate places on a street grid to a fourth location; the relational question adds the complication that the three original places are moved. For item 2, the multistructural question requires the student to find the latest train that can reach a destination by a certain time; the relational question adds the complication that there is a certain time required at each end of the train journey for walking to and from the station. The item 2 question clearly demands that the student go beyond the immediate information provided by the timetable and use the timetable information in the context of a more complicated problem. The item 1 question uses different information from that provided by the original street grid, but the new information is of the same *kind* as the original—the student is asked to construct a revised grid. This is certainly more difficult than the multistructural question, but it does not clearly involve the understanding of a *relationship* among the pieces of information in the stimulus. What might a "taxi-cab geometry" item that was relational look like? Perhaps if the students were asked to use some standard geometrical concepts in the taxi-cab geometry world, say, "What does a circle look like in taxi-cab geometry?" we might see more consistency between item 1

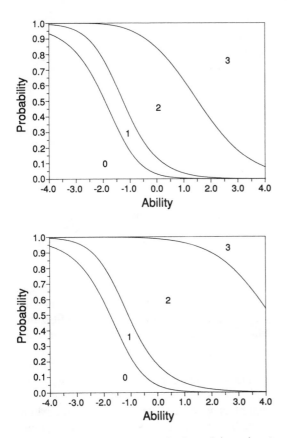

Figure 12.5. Item response maps for items 2 (upper) and 3 (lower).

and the rest.

Item 3 (see page 200) displays a divergent pattern also, but this time the relational question is more difficult than the remainder of the items. The lower panel of Figure 12.5 shows a very wide band for score 2 (multistructural), which makes a response on the relational level quite unlikely for this item. This item concerns the approximation of lengths of line segments to the nearest inch and half-inch, using a ruler. The unistructural question requires the student to estimate the length of a line segment to the nearest inch; the multistructural question asks the same question, but specifies half-inches; and the relational question makes this harder by misaligning the line interval with the end of the ruler and fails to specify the standard (i.e., inch or half-inch). Given this description, the distinction between the uni- and multistructural questions does not appear to fit as well into the SOLO framework. The relational question is obviously going to be harder for students, but this time, it seems that the

inconsistencies between this item and the others may be confusing students. This may be causing the relational question to appear very difficult.

Items 4 and 5 (see pages 201–202) display a similar pattern to item 2. Item 4 concerns a survey of people attending a football game and item 5 concerns the proportional mixing of liquids. As these two items, along with item 2, constitute the most generally consistent block of items, they will not be discussed at this point.

Items 6 (top panel of Figure 12.6) and 7 (lower panel of Figure 12.6) exhibit a quite different pattern of Thurstone thresholds from that of items 2, 4, and 5. For both, the unistructural threshold is much more difficult than that of the other items, and the multistructural threshold is correspondingly harder also. This has resulted in item response maps that are "pushed" to the right compared with those for the other items. Item 6 (see page 202) has been criticized elsewhere (Romberg,

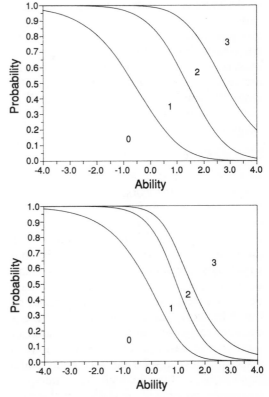

Figure 12.6. Item response maps for items 6 (upper) and 7 (lower).

1982; Wilson & Iventosch, 1988) on the basis of an ambiguous and relatively complicated multistructural question, so this issue will not be pursued here. One would not expect this problem to make the unistructural question unusually difficult — it is a seemingly straightforward graph-reading question, although it does use the word "average," which might mislead some students into trying to calculate a mean. Item 7 (see page 203) is a probability question about guessing the month and season in which people's birthdates fall. Given some familiarity with probability, the unistructural question seems a straightforward estimation of an expected value. Perhaps the explanation of the discrepancy here lies not in possible misapplication of the SOLO taxonomy, but rather in the lack of familiarity of students in the sample with statistics and probability. This would explain the translation of the Thurstone thresholds for the uni- and multistructural questions towards the difficult end of the scale. The relational questions in both cases do not experience as great a shift. This might indicate that the lack of familiarity of the more able students with statistics and probability was less marked than that of the less able. This may be due to such topics being customarily included in enrichment portions of curricula, or, possibly, that students who are more able in general have sufficient mathematical intuition and attention to detail to succeed on these items, where less able students need instructional exposure.

Fit of the items

The fit of the items, as indexed by the Item Fit t, indicates that the worst case, by a considerable degree, is that of item 1 (t = 5.78). The origin of this lack of fit can be examined by considering the empirical item response map (solid lines in Figure 12.7), which is constructed by calculating the proportions of students at each total score that make each item score, and then plotting these proportions on the ability metric, as was done for the theoretical item response maps. An examination of this figure reveals two "blips" in the empirical map: one between −2.0 and −1.0 logits, and a smaller one at about 1.0 logit. Some perspective on the meaning of "deviation" in this case can be gained by superimposing the estimated item response map on the empirical one. The

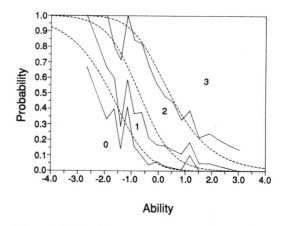

Figure 12.7. Empirical and theoretical item response maps for item 1.

dashed lines in Figure 12.7 show that the theoretical response curves are very discrepant at the lower end, but tend to fit somewhat better at the top end, other than the second "blip." Notice how the theoretical curves tend to balance between over- and underestimating the empirical curves above −1.0 logits. In calculating this statistic, greater weight is given to parts of the scale where more information is available, so it is not necessarily the case that the most important contributors to the statistic are the discrepancies that look largest on Figure 12.7.

This analysis indicates that Item 1 is susceptible to some sort of misinterpretation by students of lower abilities. The estimated step difficulties and thresholds are being determined mainly by the behavior of students of ability greater than −1.0 logits. The scores of students of lower abilities are being somewhat overestimated by these values. It seems as if some confusion occurs in students near −1.0 logits that makes the questions relatively harder. Perhaps students who recognize the grid as being a Cartesian coordinate system make the problem harder for themselves by trying to solve for Euclidean distances. This is the sort of issue that can only be unravelled by gathering more information from students about their problem-solving tactics.

As a comparison, Figure 12.8 shows the theoretical and empirical item response maps for item 3, which had a much better Item Fit (t = .50). For this item, the considerable discrepancy at the lower end has not had so great an effect on the discrepancies at the upper

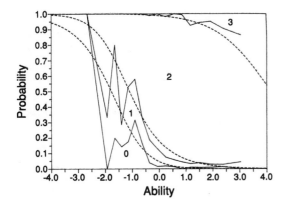

Figure 12.8. Empirical and theoretical item response maps for item 3.

end, where most of the weight of the student information lies. This situation raises the possibility of an alternative interpretation of the large fit statistic for item 1. Perhaps the students at the lower end are simply less consistent about their problem solving than those over −1.0 logit, and the sum of these inconsistent responses for item 1 was one that, by chance, affected the estimation procedure. Unfortunately, given the data, there is no way to determine the most likely of these possibilities. The empirical results and the analysis of them using an IRT model can show inconsistencies, but interpretation of such inconsistencies must be accomplished by probing more deeply into the students' cognitions than is revealed by scores on the items.

The "Doing Mathematics" variable

One question that might be asked about these items is whether attitude towards mathematics could be associated with differing item response maps. Note that as the item response maps are conditional on ability, the question being asked is not whether students who are more positive towards mathematics succeed more often, but whether their problem-solving ability differs in structure from those who have a less positive attitude towards mathematics. Given the data available, this question can be addressed only in the context of the seven items, but the technique is generally applicable and seems worth including here for illustrative purposes.

The items on the Doing Mathematics subscale of the mathematics questionnaire were analyzed using the PCM and found to give reasonably consistent estimates. The items, which used a five point Likert response scale, are shown in Table 12.1, along with the direction in which they were scored. The students were then divided into approximate upper quartile (most positive, N = 351) and lower quartile (least positive, N = 350) groups according to the scale. The responses of each group were then analyzed using the PCM.

Using the standardized difference statistic, the step difficulties that reached significance were those for steps 2 and 3 in item 2, and step 3 in item 7. For both analyses, the pattern of item fit results was the same as for the analysis of the full data set, but the values were somewhat less extreme. The theoretical item response maps for these two items are shown in Figure 12.9 (item 2 in the upper panel and item 7 in the lower panel). In this figure, the solid lines indicate the curves for the upper quartile, and the dashed lines indicate the curves for the lower quartile. The numbers in the item response maps indicate

Table 12.1. "Doing mathematics" items.

Number	Text	Orientation
1	It means doing something basic, which is the key to everything else.	Positive
2	It does not mean anything, it is nonsense.	Negative
3	It is doing something that you are told to do and that you have to keep doing over and over like a machine.	Negative
4	It is doing something which I think I just can't do.	Negative
5	It is constantly discovering something new.	Positive
6	It is doing something required, something you have to do.	Negative
7	It is a way of training my mind.	Positive
8	It is trying to find connections between different things.	Positive

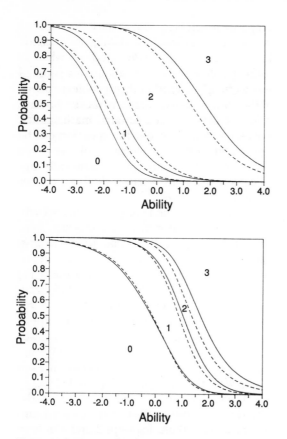

Figure 12.9. Theoretical item response maps for items 2 (upper) and 7 (lower) for two groups.

made the students who like doing mathematics less likely to make the transition to the relational level. Perhaps it is associated with the problem discussed earlier concerning students who confuse the street grid with another mathematical context (and one might speculate that it would be the students who like doing mathematics who would know about other contexts). For item 7 (lower panel in Figure 12.9), the discrepancy is clearly associated with the transition to the relational level, and the difference is in the same direction as that for item 2—students who like doing mathematics more are finding the transition relatively harder than those who like doing mathematics less. These results give one pause. It seems less than ideal to find that liking mathematics is negatively associated with success on higher levels of two of the items. Perhaps the Doing Mathematics scale is actually measuring a "reality" variable—students who score lower are simply being realistic about the difficulty of the mathematics with which they are faced, while students who score higher have a tendency towards overconfidence. In line with this, perhaps the upper levels of the items are measuring diligence rather than higher levels of mathematical problem-solving ability.

Discussion

The results described above have pointed out some specific problems with particular items. Such results are not very useful if instrument design is seen as a hit-or-miss, once-only task. If, however, instrument design is seen as an incremental process, involving the gathering of information at various times in a variety of contexts, then the lessons learned from this analysis may be put to some good use. The empirical results can be used to sharpen the technique of translating the SOLO scheme into the reality of mathematical problem-solving items. The need to sharpen the distinction between multistructural and relational for one of the items was noted. Another item needs closer examination to clarify why its relational level is so difficult. One item displayed some inconsistency that may be due to confusion that it caused for certain students. The best

the scores and are located in regions common to both sets of curves. In comparing the upper panel in Figure 12.9 to Figure 12.5, note that the curves for the full group fall between the curves for the two subgroups, a not unusual finding. For item 2, it seems that, conditional on ability, being more positive about mathematics is associated with finding it relatively less difficult to make the transitions to the uni- and multistructural levels, but relatively more difficult to make the transition to the relational level. This is what makes the band for score 2 so much wider for the students who are more positive towards mathematics.

What might this mean? If the shift for the more positive students had been consistently towards the easier end of the difficulty spectrum, we might have interpreted this to mean that success on this item is relatively more likely for them. This might have been an expected outcome, and was in fact the case for the first two transitions. But something has way to explore such empirical results is to collect samples of qualitative data at the same time as the item scores are recorded. This could be as straightforward as collecting a sample of the students' answer sheets (especially if they were encouraged to "show their work"). A more formal strategy would be to interview a sample of the students taking the test.

An interesting and informative way to look at SOLO data of the kind described here is to analyze both the dichotomous question-level data and the polytomous SOLO super-items using IRT models, then compare the two (Wilson, 1988; Wilson & Iventosch, 1988). Such a comparison could not be done in this case because the only data available for this analysis were the superitem scores.

Comparing the SOLO data with the Doing Mathematics variable seemed to raise some doubts about the relational level. Looking more closely at some of the relational questions within the items (e.g., items 1, 2, 3, and 4) leads one to speculate whether the relational level has been well-realized by these items. Certainly the relational question within each of these items would be expected to be more difficult, but that is not sufficient for it to be considered as indicating a higher level within the SOLO taxonomy. For example, in item 2, the relational question asks the student to place the use of a railway timetable into the broader context of a real-life problem where one has to consider time taken to get to and from the railway station. This is adding an extra *variable* to the problem, but is it addressing the *mathematical relations* among the components of the timetable? What is needed is a strongly mathematical idea of how to apply SOLO. One potential source for this is the van Hiele (1986) mathematics learning sequence. If one compares the SOLO idea, which is a general approach, to the van Hiele approach, one realizes that the van Hiele levels constitute successive relational levels that could be used in a SOLO framework. The interesting complication is that SOLO provides a framework for assessing *within* the van Hiele levels, and van Hiele levels provide a framework for linking *between* SOLO items at different levels.

Acknowledgments

This study was sponsored by the National Center for Research in Mathematical Sciences Education, School of Education, University of Wisconsin-Madison. The author would like to thank Tom Romberg, Director of the Center, for providing the data on which this study is based and for his encouragement and support.

References

Aitchison, J., & Silvey, S.D. (1957). The generalization of probit analysis to the case of multiple responses. *Biometrika, 44,* 131-140.

Andrich, D. (1985). A latent-trait model for items with response dependencies: Implications for test construction and analysis. In S.E. Embretson (Ed.), *Test design: Developments in psychology and psychometrics.* Orlando, FL: Academic Press.

Biggs, J.B., & Collis, K.F. (1982). *Evaluating the quality of learning: The SOLO Taxonomy.* New York: Academic Press.

Carpenter, T.P., & Moser, J.M. (1984). The acquisition of addition and subtraction concepts in grades one through three. *Journal for Research in Mathematics Education, 15,* 179-202.

Collis, K. (1983). Development of a group test of mathematical understanding using superitem SOLO technique. *Journal of Science and Mathematics Education in South East Asia, 6*(1), 5-14.

Collis, K.F., & Davey, H.A. (1986). A technique for evaluating skills in high school science. *Journal of Research in Science Teaching, 23*(7), 651-663.

Cureton, E.E. (1965). Reliability and validity: Basic assumptions and experimental designs. *Educational and Psychological Measurement, 25,* 326-346.

Edwards, A.L., & Thurstone, L.L. (1952). An internal consistency check for scale values determined by the method of successive integers. *Psychometrika, 17,* 169-180.

Glaser, R., Lesgold, A., & Lajoie, S. (1987). Toward a cognitive theory for the measurement of achievement. In R. Ronning, J. Glover, J. Conoley, & J. Witt (Eds.), *The influence of cognitive psychology on testing.* Hillsdale, NJ: Lawrence Erlbaum Associates.

Guttman, L. (1941) The quantification of a class of attributes: A theory and method for scale construction. In P. Horst (Ed.), *The Predication of Personal Adjustment,* pp. 319-348. New York: Social Science Research Council.

Masters, G.N. (1982). A Rasch model for partial credit scoring. *Psychometrika, 47,* 149-174.

Masters, G.N., & Wilson, M. (1988a). *Understanding and using partial credit analysis: An IRT method for ordered response categories.* Melbourne: University of Melbourne, Centre for the Study of Higher Education.

Masters, G.N., & Wilson, M. (1988b). *PC-CREDIT* (computer program). Melbourne: University of Melbourne, Centre for the Study of Higher Education.

Rasch, G. (1960). *Probabilistic models for some intelligence and attainment tests.* Copenhagen: Denmarks Paedagogiske Institut. (Reprinted. University of Chicago Press, 1980).

Romberg, T.A. (1982). *The development and validation of a set of mathematical problem-solving superitems.* (Executive summary of the NIE/ECS Item Development Project.) Madison, WI: Wisconsin Center for Educational Research.

Romberg, T.A., Zarinnia, E.A., & Collis, K.F. (1989). A new world view of assessment in mathematics. In G. Kulm (Ed.), *Assessing higher order thinking in mathematics.* Washington, DC: American Association for the Advancement of Science (chapter 2, this volume).

Romberg, T.A., Collis, K.F., Donovan, B.F., Buchanan, A.E., & Romberg, M.N. (1982). *The development of mathematical problem solving superitems.* (Report of NIE/ECS Item Development Project.) Madison, WI: Wisconsin Center for Education Research.

Romberg, T.A., Jurdak, M.E., Collis, K.F. & Buchanan, A.E. (1982). *Construct validity of a set of mathematical superitems.* (Report of NIE/ECS Item Development Project) Madison, WI: Wisconsin Center for Education Research.

Samejima, F. (1969). Estimation of latent ability using a response pattern of graded scores. *Psychometrika, Monograph Supplement No. 17.*

Urban, F.M. (1908). *The application of statistical methods to the problems of psychophysics.* Philadelphia: The Psychological Clinic Press.

Van Hiele, P.M. (1986). *Structure and insight: A theory of mathematics education.* Orlando, FL: Academic Press.

Webb, N.L., Day, R., & Romberg, T.A. (1988). *Evaluation of the use of "Exploring Data" and "Exploring Probability."* Madison, WI: Wisconsin Center for Education Research.

Wilson, M. (1988). Detecting and interpreting local item dependence using a family of Rasch models. *Applied Psychological Measurement, 12*(4), 353-364.

Wilson, M., & Iventosch, L. (1988). Using the Partial Credit model to investigate responses to structured subtests. *Applied Measurement in Education, 1*(4), 319-334.

Wright, B.D., & Masters, G.N. (1982). *Rating scale analysis.* Chicago: MESA Press.

Wright, B.D., & Stone, M. (1979). *Best test design.* Chicago: MESA Press.

Appendix to Chapter 12

1. The lines on the graph are city streets. One-way streets for vehicles are indicated by arrows.

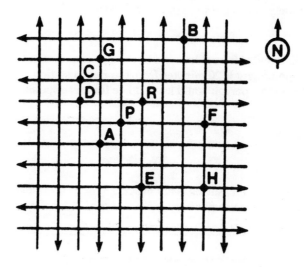

A. How many blocks would Alice (A) have to walk to visit her friend, Gayle, who lives at G, if she walks by the shortest way?

Answer _____

B. Alice (A) and Bill (B) have a friend Clara who lives at C. The three of them are walking from their homes to meet at a restaurant (R). Who has the furthest to walk?

Answer _____

C. If Bill (B) moves 2 blocks east and 5 blocks south, Gayle (G) moves 4 blocks south and 2 blocks west, and Alice (A) moves 6 blocks east and 2 blocks south, which person now has the farthest to go to the restaurant by car if the car takes the shortest possible route from each home?

Answer _____

2. A train leaves Alma and arrives in Balma at these times in the summer:

Leave Alma	Arrive Balma	Leave Alma	Arrive Balma
6:05 a.m.	6:50 a.m.	11:35	12:20 p.m.
6:55	7:40	2:08 p.m.	2:53
7:23	8:12	3:35	4:20
7:42	8:17	4:50	5:30
8:03	8:43	5:12	5:47
9:20	10:05	5:34	6:14
10:35	11:20	7:35	8:20

A. What is the latest train from Alma you can get if you want to
reach Balma by 4:30 p.m.?

Answer _____

B. If you are busy working all morning and cannot travel before 10:00 a.m.,
what is the latest train you can get so as to reach Balma by 3:00 p.m.?

Answer _____

C. A person lives 30 minutes from Alma and has an appointment in Balma
at 1:30 p.m. The appointment is 20 minutes from the Balma station.
What is the latest time this person could leave home for this appointment?

Answer _____

3. When we use a ruler our measuring is not exact. To the nearest inch, the lines below are each
3 inches long. The lengths are somewhere in the range of 2½ inches to 3½ inches.

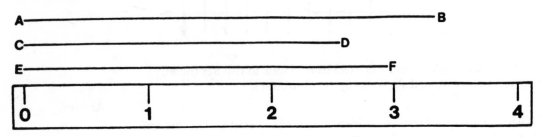

A. What is the length, to the nearest inch, of the line *EF*?

Answer _____

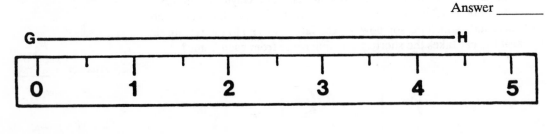

B. What is the length of $\overline{GH}$

Answer _____

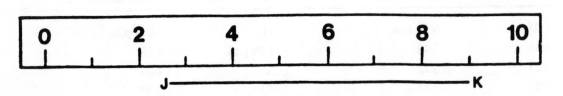

C. What are the smallest and largest possible lengths of *JK*?

Answer _____

4. A survey was made of people going into a football stadium. It was found that most people had season tickets. Only 10 people in every 100 paid the general admission charge of $2 at the gate.

A meter was used to count the people as they entered the stadium. At one time the meter looked like this:

This meter tells us that at that time six thousand three hundred ninety-two people had entered the stadium.

| 0 | 6 | 3 | 9 | 2 |

A. If 5 people went in after the meter showed 06392, what would it then show?

Answer _____

B. The attendance for games on the first five Saturdays of the football season were

Game	1	2	3	4	5
Attendance	06021	07358	10211	06102	06940

Arrange the games in order of attendance size beginning with the *smallest*.

Answer _____

C. If the meter showed 03400, how much money would be collected at the gate?

Answer _____

5. John makes orange juice by mixing orange powder with water. He measures the powder and the water in little cups, empties them in a jug, and mixes them. In the picture below he is mixing 2 cups of orange with 1 cup of water in jug A and 1 cup of orange with 2 cups of water in jug B.

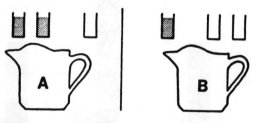

A. Which jug below would taste more strongly of orange or would they both taste the same?

Answer _____

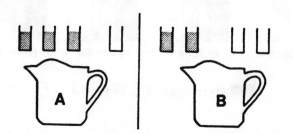

B. Which jug below would taste more strongly of orange or would they both taste the same?

Answer _____

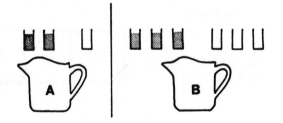

C. Which jug below would taste more strongly or would they both taste the same?

Answer _____

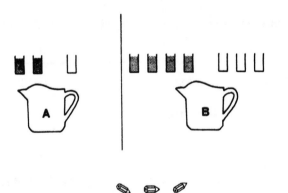

6. The figure below shows the average birth rates, marriage rates, and divorce rates in Mapland for each 10-year period beginning in 1925 up to 1974.

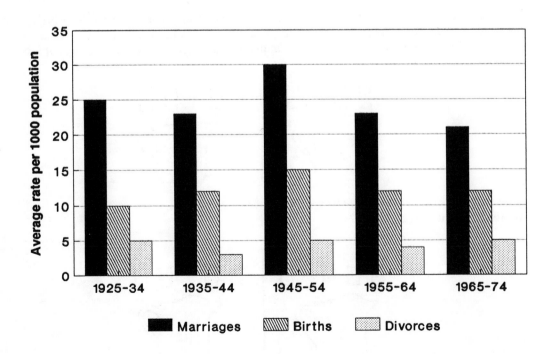

A. What was the average marriage rate in the years from 1925 to 1934?

Answer _____

B. Between which two periods did the average marriage rate decrease while the average birth rate increased?

Answer _____

C.What relationship seems to exist in general between birth rate and marriage rate?

Answer _____

7. A teacher tries to guess the season and month when any child in her class was born. If the teacher was *to guess the season*, she would most likely get 1 correct for every 4 guesses.

If the teacher was to *guess which month* any child was born, she would be likely to get 1 correct for every 12 guesses.

A. If the teacher used the *seasons* to make her guesses, how many times do you think she would have been correct with four children's birthdays?

Answer _____

B. The teacher has 12 girls and 16 boys in her class. She guessed the month in which each girl was born and the season in which each boy was born. In how many of her 28 guesses was she likely to have been correct?

Answer _____

C. If the teacher guessed 7 right out of 16 for the seasons and 6 right out of 12 for the months, how many more correct guesses altogether has she made than you would expect by chance?

Answer _____

This is the end of the questions. If time remains, you may go back and check your work or complete questions you haven't answered.

Eva Baker
University of California–Los Angeles
Los Angeles, California

Paul Cobb
Purdue University
West Lafayette, Indiana

Kevin F. Collis
University of Tasmania
Hobart, Tasmania, Australia

Joseph Faletti
Educational Testing Service
Princeton, New Jersey

Diana Lambdin Kroll
Indiana University
Bloomington, Indiana

Gerald Kulm
American Association for the
    Advancement of Science
Washington, D.C.

Richard Lesh
Educational Testing Service
Princeton, New Jersey

Frank Lester, Jr.
Indiana University
Bloomington, Indiana

Joseph I. Lipson
California State University–Chico
Chico, California

Sandra P. Marshall
San Diego State University
San Diego, California

Michael E. Martinez
Educational Testing Service
Princeton, New Jersey

Curtis C. McKnight
University of Oklahoma
Norman, Oklahoma

John G. Nicholls
The University of Illinois–Chicago
Chicago, Illinois

Tej Pandey
California Department of Education
Sacramento, California

Thomas A. Romberg
University of Wisconsin
Madison, Wisconsin

Dorothy Strong
Chicago Public Schools
Chicago, Illinois

Grayson Wheatley
Florida State University
Tallahassee, Florida

Mark Wilson
University of California–Berkeley
Berkeley, California

Terry Wood
Purdue University
West Lafayette, Indiana

Erna Yackel
Purdue University–Calumet
Hammond, Indiana

E. Anne Zarinnia
University of Wisconsin–Whitewater
Whitewater, Wisconsin